Humans before Humanity

To my parents, Jean and Nelson Foley

Humans before Humanity

Robert Foley

Copyright © Robert Foley 1995

The right of Robert Foley to be identified as author of this work has been asserted in accordance with the Copyright and Patents Act 1988.

First published 1995

Blackwell Publishers Ltd
108 Cowley Road
Oxford OX4 1JF
UK

Blackwell Publishers Inc.
238 Main Street
Cambridge, Massachusetts 02142
USA

British Library Cataloguing in Publication Data

A CIP catalogue record for this book is available from the British Library.

Library of Congress Cataloging-in-Publication Data

Foley, Robert
  Humans before humanity / Robert Foley
    p.   cm.
  Includes bibliographical references and index.
  ISBN 0–631–17087–1
  1. Human evolution. I. Title.
GN281.F66   1995
573.2—dc20   95–21626
                                                  CIP

Typeset in 10 on 12½pt Palatino
by Grahame & Grahame Editorial, Brighton, East Sussex

Printed in Great Britain by
T. J. Press Limited, Padstow, Cornwall
This book is printed on acid-free paper.

Contents

Preface   vi

Acknowledgements   ix

1 A Question of Evolution   1
2 Why Darwinism?   14
3 What are Human Beings?   32
4 When did we Become Human?   49
5 Was Human Evolution Progressive?   80
6 Why Africa?   105
7 Is Human Evolution Adaptive?   132
8 Why are Humans so Rare in Evolution?   150
9 Why Did Humans Evolve?   172
10 Does Human Evolution Matter?   195

Notes   215
References   222
Index   233

Preface

There are two functions a preface can serve, neither of them entirely honourable. One is that it is read by one's friends and colleagues to see if they are mentioned in the acknowledgements. They should skip to the end. The other function is to say what the book is about. For many life is too short to read a whole book, and therefore the most useful thing a preface can do is to help them with this dilemma.

In brief this book sets out to understand why humans evolved and what the implications of this may be. There is, however, no grand theory or simple explanation based on some key way in which our ancestors departed from the norm of evolution. Neither is there any startling catastrophe discovered here, no continents sinking into the sea or visitors from outer space. This book, within the genre of 'where do humans fit in', is revolutionary in proposing that out of very ordinary evolutionary mechanisms, monumental outcomes can occur. I believe that humans evolved because of the specific circumstances, in time and in space, in which ancient populations found themselves. The grand pattern of evolution that we can see with hindsight is in fact the product of the local and the small scale. The scale of the consequences of humans having evolved do not, perhaps, match the scale of the causes. In a field where the dramatic, the romantic, the mystical and the unusual never fail to catch the imagination, and where there is strong resistance to straightforward science and the simplifying theories of evolution, the challenge is to convince that the answer lies not in the stars and the sea, but in the day-to-day lives of ancient populations of apes faced with very specific social and ecological problems.

I shall argue that to explore human evolution it is necessary to link the generalities of evolutionary theory and biology to the specifics of the times and places in the past where human evolution actually took place. Neither on their own is sufficient, as it is the interaction between the two that is the key to explaining the evolution of humans. Such an approach can provide insights into humans and their place in nature without either losing sight of their uniqueness or resorting to special pleading. If God is in the details, then evolution is in the context.

This, then, is a book about human evolution, and by no means the first. Indeed, there has been a large number of excellent books in recent years on the subject. In the light of what I have said above I cannot claim to have written a book that would compete with the glossy adventures of fossil hunting, the Indiana Jones end of the subject; neither have I written a compendium of anatomical details. Least of all do I have any flashy theories that explain how with one evolutionary bound our hero – *Homo sapiens* – was, if not free, at least well ahead of the competition. In writing I have had two aims in mind – first, to provide sufficient information about the details of human evolution to show that it is a complex series of events, but nonetheless explicable in terms of some general evolutionary principles; and secondly, to structure it around the broad types of question that scientist and layman alike feel are significant.

What I have attempted to do, therefore, is to provide some tentative answers to a number of key questions about the human species, based on our understanding of how evolution works and the evidence of what happened during the course of human evolution. Both the questions and the answers take different forms, and for those who may want to go beyond the preface and yet not have to plough through the whole book, chapters 1, 2 and 3 are of a more philosophical nature, chapters 4, 5, 6 and 7 focus on evolutionary processes and the ecology that underpins our biology and evolution, while chapters 8, 9 and 10 address the more challenging issues of human uniqueness that usually form the core interests of the human behavioural sciences – intelligence, culture, social behaviour and language.

Turning to the many debts I owe, it is usual to start these with one's colleagues and end with one's long suffering spouse. As my wife is also a colleague in whose constant debt I am for ideas, information, encouragement and criticism, this places me in something of a quandary, so a heartfelt thanks to both Dr P. C. Lee and to Pili at the outset. About all she did not do was actually write the book itself, but we can't have everything.

PREFACE

Academics spend most of their lives talking about ideas with friends, colleagues and students, exchanging the gossip and information that both drives the subject on and creates a community. It would be impossible for me even to remember all the people who have helped me, from the seminar talk which gave me a new insight to the awkward question from a first-year student. Even exam question setting meetings have been grist to the mill. Let me offer my thanks then in the first place by class. First, to the British primate biology and human evolution community, primarily at Cambridge, UCL, Liverpool, Durham and St Andrews universities, as well as at the Natural History Museum. This is a small nucleus of scientists held together only by our fierce competition for the pathetic grant available each year for research into evolution. That we all still buy each other drinks is the main thing, but through their diversity and enthusiasm they provide the context in which this work is possible, and which it is often easy to take for granted. Secondly, to my colleagues at Cambridge, and especially the Human Evolutionary Biology Research Group. The staff and research students who have worked with me over the last five or more years have provided one of the pleasantest working environments, as well as a very important stimulus to new ideas and methods. Research supervisors are supposed to lead from the front, but these days I feel more like the reluctant officer being dragged over the top. Thirdly, the University of Cambridge and King's College have provided both financial and other forms of support.

Finally, to name some names, I thank Louise Humphrey for reading and commenting on an earlier draft, and Maggie Bellatti for helping with the figures. Nick Mascie-Taylor has been a most sympathetic and patient Head of Department. I owe an especially great debt to Marta Lahr for the virtually continuous discussions we have had on all aspects of human evolution over many years, and for her efforts to curtail some of my wilder ideas. I should also thank my children, Hugh and Conrad, for tolerating my enthusiasm for human evolution when everyone sensible knows that dinosaurs are much more interesting.

ROBERT FOLEY
Cambridge
August 1995

Acknowledgements

The author and publisher are grateful to the following for permission to redraw and reproduce figures and photographs.

Page 8: Frontispiece from *Charles Darwin and His World*, by J. Huxley and H. Kettlewell. Thames & Hudson (1975), by kind permission of the Hulton Picture Company.

Page 10: Picture by J. Reader, Science Photo Library.

Page 11: *Monkeys and Apes* by Alan Heatwole. Gallery Books. Picture by Erwin and Peggy Bauer.

Page 15: *Apeman* by R. Caird. Boxtree (1994), by kind permission of the Bridgeman Art Library/Bible Society, London.

Page 18: *The Aquatic Ape, Fact or Fiction*, by M. Roede et al. Souvenir Press (1991).

Page 42: *Primate Evolution*, by Glenn Conroy. W.W. Norton & Co.

Page 89: M. M. Lahr and R. A. Foley, *Evolutionary Anthropology* (1994), 3(2): 48-60.

Page 111: *Apeman*, by R. Caird. Boxtree (1994).

Page 125: Adapted from C. Stringer and C. Gamble, *In Search of the Neanderthals*. Thames & Hudson (1993).

Every effort has been made to trace all the copyright holders but if any have been inadvertently overlooked the publishers will be pleased to make the necessary arrangement at the first opportunity.

1

A Question of Evolution

The Scientific Counter-revolution

The strangest revolution in science is the Darwinian one. According to the conventional story the publication of the *Origin of Species* by Charles Darwin in 1859 led to a stormy but brief battle between religion and science, with evolution triumphing rapidly over creation to become the orthodoxy. Where before, philosophers, theologians and scientists as well as the great majority of educated people had believed in the fixed and immutable nature of creation, the short history of a divinely ordained world, and a special and unique place for humans in the universe, subsequently science and scientists represented the vanguard of a world view in which humans were just another species and the world was a shifting quicksand of competition and change. When T. H. Huxley scored his debating points over the Bishop of Oxford, he appeared to have delivered the *coup de grâce*. Depending on one's point of view this turning point represented either the end of the civilizing influence of religion and the collapse of Western Christian society, or else the final victory for rational thought over medieval superstition and the culmination of the Enlightenment.

The reality was very different. The revolution was neither swift nor final. In many ways it is still incomplete. The illusion of victory comes from the misconception that the argument lay essentially between established religion and modern scientific thought. This was not the heart of the battle, though, but merely one skirmish among many. The enemies of Darwinism were not only, or even most importantly, the established church, but many different groups. The idea of evolution by natural selection threatened several intellectual positions. It questioned the assumption of human uniqueness and human separation from the rest of the animal world, opening up an entirely different perspective on living organisms and the environment. It seemed both to reinforce existing notions about the progressive nature of human history, and to open up the possibility of a world with neither purpose nor direction. Darwinism also posed an entirely new methodology for looking at humans, a reductionist one where complex and philosophical concepts were ruthlessly rejected in favour of simplification, empirical observation and experiment.[1]

Darwinism, in other words, was offensive to nearly everyone. Opposition to evolution has come from all directions – from the political left for raising the possibility of genetic and biological determinism, from the right for undermining the traditional values of society, and from intellectuals for what they perceive to be its over-simplifications and attempts to reduce social complexity to the outcome of instinctive and selfish individual action. To some it demeaned humans by comparing them with animals, while to others it failed intellectually by lacking the *gravitas* of most philosophical ideas. As a result, the twentieth century has been replete with attacks on Darwinism, and, as John Maynard Smith has noted,[2] 'Darwin was wrong' has long been a favourite journalistic headline.

Today responses to Darwin and evolutionary ideas remain diverse, ranging from the critical, the indifferent and unbelieving, to the rampantly supportive. Criticism and disbelief come from a number of directions, most notably the fundamentalist wings of most religions. The fanatical and partisan element of this can be seen most clearly in the various attempts within the USA to limit the teaching of evolution, to label it as a 'belief' or unsupported theory, and to have creation science given the same educational time.[3] Such fundamentalist critics are the inheritors of the nineteenth-century clerics, although changes in orthodox Christian theology over the last hundred years, often to accommodate evolutionary ideas, have meant that arguments are now focused almost entirely on the issue of the literal truth of the Bible rather than the whole range of Christian belief.

Despite the survival and indeed proliferation of much fundamentalist thought, religion does not provide the main intellectual opposition to classic Darwinian thought. The growth of social theory and cultural relativism have produced ideas that are equally antagonistic towards evolution. Almost more insulting than the active opposition of the religious fundamentalists is the dismissal of evolutionary and Darwinian ideas by social science as irrelevant.

The opposition of most social scientists arises for a number of reasons. Principal among these is a historical one. Early attempts to apply Darwinian ideas to humans – and to social processes in partic- ular – by thinkers such as Herbert Spencer were both wrong-headed and scientifically inept, and the political agenda derived from evo- lution rapidly became the politically unacceptable face of science.[4] The social Darwinians, for example, placed most emphasis on social and ethnic groups as the units through which evolution occurred, giving rise to the idea that competition between groups was the pri- mary arena for the Darwinian struggle for survival. This led, by only a few short steps, to ideas of racial supremacy. In addition the cen- tral concept of race for much early evolutionary thinking, in the context of very simplistic notions of inheritance and genetics, meant that evolutionary ideas also became tied up with proposals for selec- tive breeding, eugenics and the purification of the human race and races. Furthermore, the underlying association between the process of evolution and progress gave an inherently distorted view of human diversity.[5]

All of this was the complete antithesis of most social thought and theory. Most contemporary ideas about humans, their behaviour, collective identity and place in the world, have derived from social theories, and thinkers such as Weber, Durkheim, Lévi-Strauss, and Leach[6] have been far more central to the development of any twentieth-century social scientific ideas about humans than have evolutionary thinkers. The explanation for this state of affairs derives at least partly from the fact that social theory has at its heart a view of human behaviour as being vested in groups and group identity, and hence being basically collectivist. Individuals may negotiate within groups, but it is their firm embedding within a social world that shapes the nature of any individualistic patterns. This is in con- trast to many perceptions of evolutionary ideas which place much greater emphasis on self-interest, on competition between individ- uals and groups, and on a reductionist view of the organization of human society. In addition, evolution has been largely associated with ideas of progress – the view that there is directional change both in the natural world and in human affairs. Progress may occur through a number of mechanisms, and may be measured or per- ceived in a number of ways – from small to large, simple to complex, primitive to advanced, and so on – but it would seem to be inherent in the evolutionary framework. This was a notion firmly refuted and rejected by many social scientists, especially anthropologists. Time's arrow was not discernible in the pattern of human diversity. Variation could be better understood in terms of the variety of social functions or the arbitrary nature of culture, and increasingly the whole enterprise of comparing human behaviour and societies has been extensively criticized. In general terms it has further been argued by social scientists that as humans are essentially social it is through social mechanisms and processes that their behaviour should be assessed. As the social is largely seen as the antithesis of what is biological, there has been a wholesale rejection of biological mechanisms or factors in the explanations of human behaviour, organization and diversity. This can be found at any number of levels, from the rejection of genetic and strictly hereditarian views and innate human characteristics, to the abandonment of environmental determinism.

The extent of this abandonment of evolutionary ideas cannot be overestimated. For most social theorists and philosophers of the twentieth century there has been little need to do more than pay a brief lip service to the probable fact of evolution having taken place. However, as the rubicon from biological to social, from behaviour to culture, and from determinism to free will had been crossed, then to take an evolutionary approach into the heart of the human dimension was perceived as at best misguided and at worst dangerous.

Perhaps the most striking and almost astonishing aspect of this state of intellectual affairs is that it all happened so long ago. Even before the cement had dried on Darwin's memorial in Westminster Abbey the tide was turning against evolutionary ideas.[7] At the turn of this century the American anthropologist Franz Boas led the way by showing how culture and race could be detached, and thus how cultural change did not depend upon any biological or evolutionary ideas. His ideas, as well as those of his contemporaries such as Malinowski, had largely eradicated an evolutionary approach by the 1920s and 1930s. In its place such influential figures in modern anthropology as Evans-Pritchard and Fortes established the timeless and culture-bound nature of human existence that has dominated the entire subject. Add to this the abhorrence associated with some so-called evolutionary ideas at the end of the Second World War, and by 1950 evolutionary anthropology was essentially dead.[8]

It is important to note how long ago that was. The rejection of evolution took place at a time when biologists were themselves only just coming to terms with the nature of evolutionary theory, before the nature of inheritance was widely understood, and at a time when the mechanism of natural selection seemed to be logically unsound. Given the disillusion with Darwinism felt by large numbers of biologists, it was hardly surprising that anthropologists and social scientists should have been so critical. However, the world of biology and evolution has changed out of all recognition in the intervening years, and yet much of the understanding within anthropology harks back to these early errors. Since Boas' time, genetics has been revolutionized by population mathematics, the nature of DNA has been discovered and its complex mechanism gradually unravelled, evolutionary theory has been completely overhauled, and a range of broader biological ideas has been linked to evolution – ecology, development, behaviour. The evolution that was rejected in the first part of the twentieth century is an entirely different discipline from the one that has emerged in the second half, but this has seldom been addressed.

It might be thought that the gradual accumulation of new facts might alter this situation, but this possibility has been greatly reduced by a new strand entering the argument – cultural relativism and deconstructionism. According to many current thinkers, there is not and can never be any objective knowledge, only a humanly created world of words and texts. The human experience, including 'knowledge', can only be filtered through the linguistic world of thought and communication. The world we experience is therefore nothing more than a construction of our senses, and in particular the language we use to describe our experiences. It is impossible to break the circle between the human world and the human experience of that world. At its most extreme, there is nothing out there except a sea of words, although a more moderate version would be that while there may be a 'real world', we can never really observe or experience it 'objectively'. Furthermore, the abundance and variety of cultural beliefs and systems emphasize this lack of any real autonomous world, and hence reduce evolutionary biology to just another culture-bound model of the world.

Evolution in this paradigm is just one more text, with no more authority than any other. Indeed the attempt to give authority, to suggest that it derives from a scientific basis, is sometimes used as evidence that it is closely associated with various entrenched and empowered establishments – usually, in no particular order, men, capitalism and the Western Anglo-American world. In this world of complete relativism evolution should be reduced to nothing more than one among many creation myths, although occasionally it may be given special status as being unusually (for a creation myth) associated with materialist values and the principles of the market place rather than the temple. And as just another text, it can be decoded endlessly as long as it is not taken seriously or as a statement of reality.

It is hard to mount an effective reply to such a criticism without recourse to a number of fairly obvious points about the direct evidence for evolution, the fact that it encompasses the totality of biological life rather than just humans and their linguistic world, and that there is mounting evidence that the view that language is entirely culture-bound is erroneous.[9] The reply to these points would obviously be that these are just further text-bound attempts to privilege evolution as a system of knowledge. Unbound by empirical fact, deconstruction can always have the last word, but word it is alone.

There is also another line of opposition to Darwinism, but this one lies within biology. The argument here is not about evolution and whether it occurred, but about how and why it took the form it did. In this sense it is largely a technical argument, but it has many echoes in some of the more philosophical debates. The crime of Darwinism is essentially the same – it is too simple an idea. Early criticisms focused on the mechanisms of inheritance, which for Darwin were largely unknown. Darwin did not have available to himself the genetic theories of Gregor Mendel, and his ideas on how offspring inherited characteristics were rather woolly and incomplete. The key issue was whether biological traits were inherited in an undiluted and particulate form, and thus swept from generation to generation without changing or being modified by the mechanisms of reproduction, or whether with sexual mating traits were blended together, and thus were in a state of constant fusion and fission. The dilemma that arose as a result, over about fifty years, provided Darwinians with a Hobson's choice. On the one hand, if inheritance involved the blending of characteristics derived from parents, then over time there would be a gradual dilution and loss of most novelty, and hence no evolutionary change as the so-called new and adaptive change would be swallowed up in the inertia of the existing forms. There would therefore be a major barrier to the selection of advantageous traits and to evolution occurring at all. On the other hand, if (as is now known to be the case) there was no blending but instead particulate inheritance from each parent, then this seemed to lead, via the process of mutation, not to gradual and adaptively led evolutionary change, but to random and rapid change. Even T. H. Huxley, Darwin's so-called bulldog, stated that natural selection was not sufficient to account for evolution, and as has been noted many times, the successive editions of the *Origin of*

Species saw a gradual watering down of Darwin's original position.[10]

While the development of what is known as neo-Darwinism or the modern evolutionary synthesis between 1930 and 1950 largely removed these objections by showing how these particulate inheritance mechanisms linked to population structure and hence to adaptation,[11] new ones have arisen within biology. There have been two main sources of criticism, from either end of the evolutionary spectrum. At the so-called 'macro' end, a number of palaeo-biologists such as S. J. Gould and Niles Eldredge[12] have proposed that evolution is not a gradual process but occurs in fits and starts (punctuated equilibrium, consisting of long periods of stasis and short bursts of change), and as a derived inference from this observation, that traditional neo-Darwinian orthodoxy cannot account for the totality of this pattern. They have thus proposed that there are additional mechanisms, and therefore that far from being simple and unitary, evolution consists of a hierarchy of mechanisms, from the smallest level of the gene or the genome up to individuals and populations and beyond to species and even whole ecological communities.[13]

At the other extreme, biologists who study living systems at the molecular level have questioned the power of natural selection. The intricacy and idiosyncrasies of the way the basic molecules are put together are thought to lead to their own mechanisms for change – for example, Gabriel Dover's molecular drive,[14] in which in effect the physical environment and structural properties of the molecule itself are more significant in driving evolutionary change than the environment of the whole organism or individual. In addition, the sheer scale of genetic material – the 30,000,000 bases that occur in the human genome – and the apparent lack of function of most of this genomic material has led others to argue that there is simply too much slack and inertia in the genetic system for evolution to be the sleek and finely tuned adaptive system that is implied by the theory of natural selection.

It is immediately obvious that the criticisms at both levels are technical versions of the criticisms from outside the realm of biology; that classic evolutionary theory is oversimplified, reductionist and takes insufficient account of the way in which collectivities, or larger entities beyond the gene, influence the process of evolution. Life is so complex, and human existence so improbable, that there must be more than just one simple process. Where technical and in-house criticism perhaps departs from the broader antagonism is that virtually all biologists are happy to accept that evolution has occurred and is significant, that historical aspects are important in the development of life, and that there is no serious problem with the apparent randomness and improbability of the evolutionary process.

Darwinian Dragon's Teeth

Given this endless renewal of criticism, it might well be thought that Darwinism is a spent force in science. This is not to say that evolution is not widely accepted, but that the formulation of it made by

Charles Darwin, author of *Origin of Species* and the founder, with Alfred Wallace, of the modern theory of evolution.

Darwin is, not surprisingly perhaps, no longer viable after nearly one hundred and fifty years, and has been replaced by a much more sophisticated and comprehensive theory or set of theories. That would certainly be the view of many authorities, and is one that has been widely expounded in both the technical and popular literature by Gould, perhaps the best known of evolutionary biologists today. He has brought the excitement of evolutionary biology to the widest possible audience, but at the same time has played a large part in relegating Darwinian theory to the status of a politically entrenched ideology and incomplete science.[15] From his position of evolutionary iconoclast he has erected a new orthodoxy, Darwinism for the politically correct. The new evolutionary theory accepts fully that evolution occurred, but emphasizes that there is no progressive direction. More importantly, the survival of species is largely the product of randomness. New structures arise largely because new structures arise. The keystone of Darwinian thought – survivorship of the individual to reproduce through some adaptive advantage, and hence the evolution of an 'adapted world' – disappears. The rather earnest and competitive evolutionary biology of the nineteenth century can be replaced by the lottery biology of the late twentieth. There is nothing to explain in human evolution because it is all a matter of luck, or of processes operating at such a high level that they do not impinge upon the minutiae of the human form.

And yet despite all this, Darwinism is like a persistent weed. Each time it is apparently eradicated, but sooner or later it always returns. Perhaps it is even more like the dragons' teeth in the labours of Hercules, growing more numerous and stronger with each re-emergence. While one view of the development of evolutionary biology this century might be the gradual retreat from orthodox Darwinism, another might be that it has sequentially triumphed in one discipline after another. Each decade there has been a new convert. For the 1930s it was population genetics,[16] and for the 1940s and 1950s natural history.[17] In the 1960s and 1970s both ecology and ethology became increasingly evolutionary and Darwinian in their approach,[18] while the 1980s and 1990s have seen the growth of a more Darwinian view of molecular biology,[19] of evolutionary psychology,[20] and even sporadic outbreaks have been found in anthropology.[21]

These changes have come about for numerous reasons. Most significant has been the slimming down of the logic of Darwinian theory (see below, p. 24) in such classic books as G. C. Williams's *Natural Selection and Adaptation*,[22] and the elegant and increasingly mathematical demonstrations of how Darwinism works and the areas into which it extends. Fieldwork and experimentation have greatly expanded the data available on the actual lives and deaths of animals, so that evolution is no longer seen as something that occurred in the past, but as a continuing process. Even the growth of the fossil record and the development of new techniques that allow more information to be extracted about life in the past have played their part. The growth of molecular biology has also shown that the Darwinian world of competition and co-operation continues unabated below the surface of the individual. All in all, Darwinism is as strong today as at any time in the past. Darwin's children are alive and well, and his theory is remarkably robust. The Darwinian revolution has proved to be durable because it cannot be reduced simply to religion versus science, the poetry of King James's bible versus incomprehensible names, and free will versus biological determinism.

Nowhere is the relevance of Darwinism greater than in considering humans and their evolutionary heritage. There have indeed been

Fossils are the main direct source of information about the evolutionary history of a species. This fossil, which is an early member of the genus *Homo* from Kenya, was reconstructed from numerous fragments.

innumerable books on this topic, and a number of well-established traditions exist. There are the books that outline the story of human evolution as an adventure in the discovery of the fossil past, with many heroic struggles and upheavals on the part of both ancestors and anthropologists. These books are replete with fossils and obscure names of human-like things that have long disappeared. There are also the books on human evolution seen through the lens of other animals living today. Desmond Morris epitomized this approach with his *Naked Ape*,[23] but in recent years there have been a number of much more empirically-based studies which have focused on detailed and long-term research on apes and monkeys in particular, showing the close kinship between humans and their relatives. These books often reveal startling capacities of animals, and then make the leap of faith across to human prehistory. There are too, of course, the many books that might be described as the lunatic fringe, books

Chimpanzees are the closest evolutionary relatives to humans alive today. A shared ancestry can be used to provide information about the characteristics of the last common ancestor, estimated to have existed over 5 million years ago.

that usually reveal the seeding of earth from outer space, in which sex plays a prominent role.[24]

Among these lines of thought a dichotomy exists. In one tradition there are endless details of fossils, tracing out a path of change, but with little sense of explanation or understanding of what could have led this to happen – all bones and no flesh. And conversely in the other, many thinkers concentrate on the broad contrasts between human and animal and the various chains of cause and effect that could lead an ape to evolve into a human, but pay only scant attention to the details of when and where this event actually took place – all flesh and no bones in a timeless past.

A central point of this book is that time and place matter. The reason why humans are here today is linked to myriad events in the past. The past, it argues, cannot just be invented or imagined, nor reconstructed solely from observations of the way the world is at the present. Equally, the bones that tell us about the past, the realm of palaeobiology, do not tell that story alone. To understand them requires the framework of the processes by which evolution works – in other words, the mechanics of Darwinian evolutionary biology. I will thus try to bring together the two themes – the evolutionary past and its evidence, because they are the time and place that are so critical, and evolutionary biology more generally because it is necessary to explain why things happened the way they did.

To this end I have adopted a particular format. Most people have an interest in human evolution because it relates to a number of questions about humans today and to human nature and history, rather than to any specific technical questions. However, the technical nature of biology and palaeontology is such that some knowledge is essential. I have therefore posed the various questions that come to mind to anyone curious about humans. In each of the chapters that follow I attempt to show how the various lines of scientific enquiry and thought can provide answers, and that these answers may have implications beyond the purely technical. In short we need both the facts of science and the conceptual framework provided by evolutionary ideas.

The Darwinian Inheritance

The framework for answering these questions, I shall argue, is provided by three of the most fruitful consequences of the Darwinian revolution – Darwin's three intellectual and scientific offspring that have survived through the various attacks and criticisms of a largely hostile world. These three are: that the question 'why humans?' is amenable to scientific enquiry through the fact of evolution; that the theory of natural selection remains the most powerful mechanism for explaining evolutionary change; and that the unravelling of the fossil past – the humans before humanity – are the best clues to a world that no longer exists, and yet of which people are the inheritors.

2

Why Darwinism?

Humans as a Scientific Problem

The acceptance of the theory of evolution is often discussed and debated in terms of the replacement of one explanation – or myth or story, depending on point of view – by another. For most Europeans brought up in the Judaic and Christian tradition this was of course the story of Genesis, the creation of the world by God and of Adam and Eve as the first man and woman. This particular cosmology accounted for both human nature and humanity's relationship to nature. Although there are often universal elements in such myths, the diversity in content is enormous, usually expressing the variation in cultural beliefs and practices. Among the Baroba of Amazonia, for example, humans were created by the anaconda as it swam upstream, vomiting each clan as it went along.[1]

The idea that Darwinism is just another story is satisfying in many ways, and is particularly popular in the context of the growth of some modern views that see all knowledge as essentially a text or a story. In some senses this may be true; the cosmology of Western science serves many of the same functions as traditional mythology. The theory of evolution places humans in a context and provides to many a basis for describing and understanding human nature. Misha Landau[2] has shown how a number of interpretations of human evolution take on the structure of classic myths and fairy tales. Such tales have a hero, a struggle, a gift and a triumph. In human evolution there is a hero who starts out from a deprived and benighted home (i.e. being an ape), who struggles against adversity (the forests disappear and the heroic hominid must battle it out on the scorching savannas), discovers or is given a gift which is the key to success (this can be anything from upright walking, to tools, to intelligence, to

At one level evolutionary theory acts as an origin myth, but it has other characteristics as well

language, depending on preference and possibly evidence), and then triumphs (i.e. becomes a modern human and acquires the trappings of civilization). This congruence between traditional and scientific stories is striking. Furthermore, the evolutionary cosmology often underpins many non-scientific, ethical and metaphysical views within current Western culture. These may range from a belief in the insignificance of humans both as a species and as individuals (after all, humans are just a minute speck in a single instant in a vast universe), to the assumption that humans are innately sexist or aggressive or racist or monogamous. Indeed, it can underpin the idea, through genetics, that there is an essence to humanity. Rather than humans being endlessly and almost infinitely reshaped by the environments in which they live, they can be claimed to have a particular nature – human nature, with the emphasis on nature.

It can thus be respectable to downplay the impact of evolutionary ideas. What has changed, if one story has been replaced by another with similar structure and purpose? Something, though, has changed. That current evolutionary models can act as a creation myth should not lead to the belief that there is not also something new, and that an evolutionary explanation can and does both go beyond the explanatory power of other myths and operate in a different way. In this context Darwin's legacy has been substantial. That legacy lies in two directions, first by making questions about why there are human beings a technical rather than a philosophical one, and secondly by placing origins in the context of the scientific method rather than in that of belief.

In the *Origin of Species* and *The Descent of Man* Charles Darwin laid out not just another narrative of human origins, but more importantly, a scientific mechanism by which humans could have arisen without the need for divine intervention. That mechanism was natural selection, and it applied equally to mice and men. Humans were not the act of a special creation but were instead merely part of a continuum of evolutionary change. Science had, in the shape of evolutionary biology, extended its grasp to the most basic of philosophical questions – why are there human beings?

Darwinism and evolutionary theory did not automatically provide an answer to that question, nor bring an end to irrational speculation. What it did do was provide a new framework in which the question could be answered. That framework is essentially naturalistic and materialistic. Humans appeared, and took the form they did, not because of any preconceived plan or grand design, but because of the interaction of particular lineages or lines of evolution with new selective pressures. Darwin's legacy was a wholly new way of asking questions about humans and their place on earth and in the universe. Needless to say this has not brought about an end to philosophical, historical or religious speculations about the human species and its place in the world. Indeed in some cases it has opened up a whole new genre of secular narrative very loosely based on evolutionary ideas or else strongly opposed to evolutionary explanations.

Rather than narrowing down the range of speculation about human origins and nature, evolutionary biology has been the spur to all sorts of fanciful ideas. Some of these have been explicitly anti-Darwinian, based on the premise that such a haphazard mechanism as natural selection could not possibly be responsible for our complexity. A favourite escape route from the constraints of biology has involved aliens from outer space. Van Daniken, in his book *The Chariots of the Gods*,[3] managed to make a great deal of money by suggesting that while evolution may have produced the run-of-the mill type of species, it took astronauts visiting Earth from another planet to sow the seeds of civilization.

Even within the eminently broad confines of Darwinism, speculation about human origins has spread far and wide. Elaine Morgan[4] has done much to popularize the idea that humans, despite being descended from arboreal primates, were in fact the product of an aquatic phase of evolution. Even more wondrously, it has even been suggested that humans are the result of miscegenation – the chance outcome of cross-species breeding between chimpanzees and orang utans.[5]

These stories of the history of our species are all tentatively evolutionary, although they bear the same sort of relationship to mainstream evolutionary biology that Bugs Bunny does to a real rabbit. They do, however, reflect a general response to Darwinian theory that is found widely in both scientific and non-scientific circles. However, Darwin's legacy, evolution by natural selection, remains as powerful a tool as when he first developed it. Essentially what the theory of evolution by natural selection does is to turn very large philosophical and metaphysical questions into what are often straightforward and even boring technical ones. The question, 'where do humans come from?' demands not a general answer but a specific one about time and place. The question 'what is unique about humans?' leads to the field of comparative anatomy and behaviour. Taken seriously, evolution does not necessarily limit the imagination, but it does require that such imagination is disciplined by such empirical facts and feasible mechanisms. This does not mean that the answers are easily obtainable or that they will be widely accepted. If there is a prosaic quality to the explanations for humanity that often

Many theories of human evolution place great emphasis on something unusual having happened in our evolutionary past. Elaine Morgan's aquatic ape hypothesis proposes that we went through an aquatic phase early in our evolution. This would be unusual for an arboreal order such as the primates.

fails to satisfy, that is the nature of modern evolutionary biology. But by asking very basic, simple and straightforward questions, very powerful, complicated and elegant answers can be found.

Science – Asking Small Questions

Peter Medawar once defined science as the art of the soluble.[6] Although this reflected the predilections of the biochemist, it also went to the heart of the scientific method – the importance of asking questions that can be answered. In Douglas Adams's *The Hitchhiker's Guide to the Galaxy* a powerful computer was built to answer the question as to the meaning of 'life, the universe and everything'. Apart from taking an inordinate time to find an answer, and finding an incomprehensible one at that (42), the question did little more than provide employment to philosophers speculating on the answer. The problem was that the question broke Medawar's rule of science – it was insoluble. In contrast to other realms of knowledge, science works best not by asking large and grandiose questions, nor by ignoring them, but by breaking them down to answerable chunks. Problems and questions should be susceptible to solution, even if that means that they are only steps to a more distant goal. For someone who wants to know the answer to life, the universe and everything, asking the question in this overwhelming form is not the way to approach it. In science, knowledge is particulate, built up piece by piece. Each piece on its own is often incomprehensible and dull, but the result is effective. Instead of enormous spans of inferential arches, there is a solid edifice. What it lacks in glamour and spectacular silhouettes it makes up for in durability.

How does this excursion into the nature of science help with the questions posed about humans? The simple point is that while all societies and cultures have asked questions about the origins of humanity, the distinctive characteristic of evolutionary approaches is this absence of grandiose questions and an often obsessive concern with details. Darwin's *The Descent of Man*, for example, is a masterpiece of apparent irrelevancies. Anyone reading a recent book on human evolution is likely to be overwhelmed by the details of anatomy gleaned from fossils. The gap between these anatomical observations and the questions that people are actually interested in – where do humans come from? – seems ridiculously large. And yet it is exactly this link that is the crux of a scientific approach to humanity and its place in the biological world. If we want to answer the questions asked at the beginning of this book, then this must be the correct approach. Although it may not appeal to those who are attracted by the exotic in science fiction or the conundrums of meta-physics, the real achievement of Darwinism has been the translation of unanswerable questions into ones which, when suitably modest, can be answered.

Human origins and ultimately human nature are not philo-sophical questions, but technical ones. The task that lies ahead is to find a way of examining the technical details without losing sight of the larger questions to which they are supposed to provide an answer. The route to this is not merely by recording the fossil evidence and the life of primates, but by placing these sets of data into a strong theoretical framework.

What sorts of questions can be asked? Obviously, given what has been discussed above, an answer to the question 'why are there human beings?' cannot be achieved by addressing the problem of the nature of the human species in this extremely general way. Rather, the big question needs to be broken down into a number of more manageable questions, the cumulative effects of which are to reveal what happened in the evolutionary past, and by implication, docu-ment the principal characteristics of humanity. Thus, for example, the question 'when did we become human?' will be posed here. While this may appear a straightforward question about the fossil record, in practice it turns out to hinge not just on the technicalities of dating fossils, but on the criteria by which humanity is defined. Answering this question in turn provides the basis for interpreting ideas about human antiquity and its effect on humans today; it makes some difference whether humanity is millions of years old, hundreds of thousands, or merely thousands. The ability to answer the relatively simple question hinges on the criteria taken to define humanity – is it language, culture, bipedalism, intelligence, tool-making or any other number of characteristics? Again, by breaking down this investigation into its component parts we will be in a position to try and understand what the function of each of these is, how it compares with what is found in the rest of the animal world, and why it might have evolved. Why, for example, do humans have large brains? To answer this question it is necessary to explore how large brains occur among birds and mammals, what advantages they provide, and why, in fact, they are not more common than they are.

This approach to science is reductionist – the attempt to explain phenomena in terms of their elemental parts and entities, with min-imal assumptions about those parts or entities. Reductionism is often contrasted with a holistic approach, one that looks at phenom-ena as a whole, as they are constituted in their entirety. Each method has its advantages and disadvantages. It is often claimed that reductionism loses the essence of the relationships between the parts as a whole. To understand the internal combustion engine, for example, little can be learnt from looking at the carburettor and the pistons in isolation. It is only the functioning whole that is actually an engine. In this sense holism may offer considerably more insight into complex phenomena. However, in practice it is often very difficult to study things as a whole. The actual processes and mechanisms involved – the movement of the pistons – can be lost in the blur of the whole. Over the years most information has been gleaned from adopting a reductionist approach at least as a working assumption. Such an approach involves the minimal number of assumptions and offers the hope, when dealing with evolutionary problems, of actually seeing how the different components fitted together in the first place. For example, with human evolution the very worst assumption possible is that the whole package – bipedalism, large brains, culture, language – all came into existence at the same time and were always articulated in the same way. By looking at these components in isolation at least there is the hope of unravelling how the characteristics of hominids and humans evolved, and how they came to take the form they did.

The Science of Evolution

There is, of course, no reason why the question 'why humans?' cannot be asked in any number of contexts, and it is extremely probable that it always will be. Intellectual life would be excessively dull if there was only one way of doing something. This book, though, is limited to asking and answering the question of why there are humans within an evolutionary framework. There is some justification for saying not that this is the only way of asking the question, but that it is the only one that can come up with some answers that are subject to empirical testing and consideration. Evolutionary theory has some primacy when it comes to humans just because it is a theory that encompasses not just humans, but the entire living world. Indeed, this is the power (and the threat) of evolutionary biology, that it attempts to put humans simply if not easily into the same framework as any other species.

The primacy of evolutionary theory comes from a number of sources. First there is the actual evidence of the fossil record, which shows that the biological world has not been stable, but has undergone major changes both in the structure of the organisms and their ecosystems, and in the taxonomic groups that exist at any one time. Although it is hard to think of the geological record as dynamic, none the less it provides the best evidence that over long periods of time there has been a pattern of continuous change. From our point of view, one implication is that there have not always been humans, and that for most (99.99999 per cent) of the time that life has been in existence on Earth it has managed perfectly well without *Homo sapiens* or even closely related forms. Evolution provides a framework for trying to understand how a world in which humans were absent became one in which they could exist and be successful. Furthermore, the fossil record, by showing the enormous diversity of life through time, provides an appropriate comparative framework for assessing what is new about humans as biological organisms, and therefore why they evolved when and where they did.

Secondly, all living matter shares the same chemical material and is built upon the same replicating molecule, the DNA (deoxyribosenucleic acid). This is evidence for the unity of life, for the fact that different forms of life did not have separate origins. It is still startling albeit now well-known that all plants and animals, single-cell and multi-cell organisms, use exactly the same genetic code. The flexibility of this molecule is itself wondrous, but rather more mundanely it provides further evidence for the fact of evolution as well as for one of the mechanisms by which it occurs – inheritance. Genes are essentially the means of transmitting information (for building new individuals) from one generation to another. The fidelity of the copying mechanism that lies at the heart of the DNA structure is the basis on which continuity and similarity in life forms is maintained, while the occasional errors that occur form the basis for the introduction of novelty – the necessary components for new species.

Most importantly, perhaps, evolution is not just a process or event occurring through time, but is the outcome of a workable mechanism by which that change occurs. It is often mistakenly thought that Charles Darwin's great contribution was the discovery of evolution. This is not the case. The idea of, and even evidence for, evolution was well known during the early nineteenth century and even before.[7] The simple concept that the world and its biological systems have not remained stable is an obvious and attractive one. The problem was not the fact of evolution, but the mechanism that could drive it. It was this that concerned the biologists and geologists who were Darwin's predecessors and contemporaries. Many mechanisms had been suggested, including the best known alternative to Darwinism, Lamarckism.

Lamarck was a biologist and came to the conclusion that evolution had occurred for much the same reasons that convinced

Darwin. Lamarck thought that organisms changed because they had an inbuilt drive to change, a need to improve themselves. This meant that things that an individual learnt or acquired during its lifetime which enhanced its survivorship would be passed on to the next generation. This was the concept of the inheritance of acquired characteristics. As an evolutionary mechanism it failed though, because it very soon became apparent that there was no mechanism by which such new information could actually be transmitted. Even more problematic was the fact that during an individual's life an enormous amount would be acquired. Many of these acquisitions might be beneficial, such as greater dexterity or even wisdom, but many were far from advantageous. The older you get the more your teeth fall out, or you develop arthritis or cancer or any number of degenerative diseases. Would offspring born after these had developed in the parents acquire these characteristics as well? There was, in other words, no sorting mechanism and no true means of inheritance.[8]

Darwin's great contribution was to provide an appropriate mechanism, what he called natural selection. Like Lamarck's theory it used the idea of descent, with modification based on some means by which characteristics increased the probability of survival. Darwin, though, added the key component of reproductive success, or number of offspring left, and omitted any role for the inheritance of acquired characteristics. (It must be admitted that under the force of contemporary criticisms arising from the fact that Darwin did not have the exact nature of inheritance worked out, he became more and more Lamarckian with each successive edition of his book.)[9] That mechanism not only convinced the nascent evolutionary biologists, but also developed into an enormously powerful tool not just for investigating the grand pattern of evolution over long periods of time, but also for understanding the details of animal behaviour and adaptation on a small scale. Although evolutionary biology today is very different from the subject that flourished in the late nineteenth and early twentieth centuries, concerned primarily with origins and relationships, with a ladder of progress and the characterization of anatomical structures, it not only still depends strongly on the fundamental Darwinian idea of natural selection, it is also probably closer to Darwin's original intent than much biology that lies in the intervening years.

However, many criticisms have been levelled at Darwinian evolution. One is that it is often described as untestable. This view was held by, among others, Karl Popper,[10] and allowed other scientists to treat evolutionary theory with a certain smugness if not contempt.

The reasoning is as follows. Evolution is a historical event; it has already occurred far back in the past, and therefore it has not been subject to direct observation. Without direct observation it is not possible to test the fact of evolution. This is certainly a major problem, but the testability of evolution does not lie with the fossil record. Evolution can be taken to mean two different things, and as we shall see below, the distinction is important. It is the process through time, and it is also the specific mechanism that causes that change. This mechanism can be tested and indeed frequently has been, and there is therefore no reason to say that the theory is untestable. The fact that evolutionary biology must perforce be a historical science is certainly a constraint, but it is a constraint shared with astronomy and cosmology. It is impossible to observe the formation of stars or planets directly, let alone the origins of the solar system and the universe itself, but none the less it is possible for scientists to investigate these events and to test stringent scientific theories.

A second criticism often levelled at the scientific status of evolutionary biology is that it is based on a tautology – that is, if evolution is the survival of the fittest, then who survives but the fit and who is fit other than the survivors. It is perhaps unfortunate that what amounts to a slogan designed to simplify an elegant theory has come to be the full representation of that theory. The slogan, taken independently, is tautological, but it omits the precise logic of the Darwinian formulation that makes independent the 'survivors' and the 'fit', or at least puts them into context. In practice the theory of natural selection is empirically testable and built up from a series of concrete observations.

Natural selection simply means differential reproductive success; that is, that given a population of reproducing organisms, and given that individuals within that population have different numbers of offspring, then natural selection is the mechanism that determines this differential reproductive and survivorship rate. This mechanism lies at the heart of evolutionary theory. In a sense, it is the bit of the theory that does all the work, sorting out the individuals in each generation and thus determining the direction of evolutionary change. Any theory of evolution would have to have some such mechanism, and simplicity is its strength. It is, though, in this simplistic form, just an assertion. For natural selection to operate, certain conditions have to be fulfilled.

The first of these is that organisms reproduce. If there were no reproduction, then the game of life would have to start afresh with each deceased generation. In the early days of life-forms perhaps this happened, when the replicating molecules were less efficient, although natural selection would still be operating, selecting for the most efficient replicating molecule.

The second condition is that there should be some mode of inheritance – that is, offspring should resemble their parents more than they do the population as a whole. This is the field of genetics, but when Darwin wrote the *Origin of Species* it was this aspect that was least well understood and caused him the most critical problems. If information that determines the characteristics of the parent can be transmitted to offspring, then those features which enhance the survival and reproductive potential of the parents will occur more frequently in each subsequent generation, dependent upon the number of offspring. If there was no mode of inheritance, then the advantageous features of a parent would simply be lost in each generation. There could be no evolutionary change. It was the absence of this condition that made Lamarckism an unworkable theory of evolution.

There must, thirdly, be variation within the population. Even if the first two conditions are fulfilled, if each individual in the population is phenotypically and genetically identical, then natural selection cannot operate. Differential survival will have no effect because all individuals are the same, and so each generation will be identical. It is for this reason that Darwin was himself so concerned with the problem of variation, and why he devoted the first two chapters of the *Origin of Species* to 'Variation under Domestication' and 'Variation under Nature'. This was in fact also one of the principal lines of evidence employed against the theory of a special creation of immutable types – if God had created a number of types of plant and animal, then unless he or she was incompetent there was no reason why they should vary at all. Darwin went to extraordinary lengths to show that even the humblest type of creature displayed variation.

Finally – and this condition is perhaps the one most closely associated with Darwin's own ideas – there is competition. Imagine a world in which all the conditions outlined above were fulfilled. However, if the resources needed to support all the populations were infinite, then there would be no differential reproduction. An individual could have all the offspring possible, and so there would be no change from one generation to another, just a constant and everlasting expansion. Clearly, though, such a world does not exist. Indeed it is theoretically impossible, as time itself is a resource (time to have offspring, etc.) and so as long as there is time, there will be at least some limitation. In practice, of course, all resources are limited – energy, water, shelter, potential mates and so on. It was the eighteenth-century demographer Malthus who first pointed out the imbalance between the potential of resources to expand and the potential of populations to reproduce. Darwin harnessed this notion as the central condition necessary for natural selection to operate. If resources are limited, then not all individuals will survive and reproduce, or they will reproduce at different rates of success. Given the conditions of reproduction, variation and inheritance, then those individuals who are better adapted to acquiring the resources necessary to survive and reproduce will leave more offspring, and those offspring will carry the feature of their parents that gave them this competitive edge.

These then are the conditions under which natural selection must occur. Each of them is independently derived and each of them is easily and empirically tested. Organisms can be observed reproducing, the mechanisms of inheritance have been worked out, the occurrence of variation can be and largely is extremely well established, and the finite nature of the resources of the world is virtually a truism. Looking at the theory of natural selection in this way shows that far from being untestable, it is in fact a logical necessity deriving from a number of simple observations.

If these conditions occur, then evolution must be a consequence. It is useful to distinguish in this way between evolution and natural selection. Natural selection is the mechanism of change, dependent

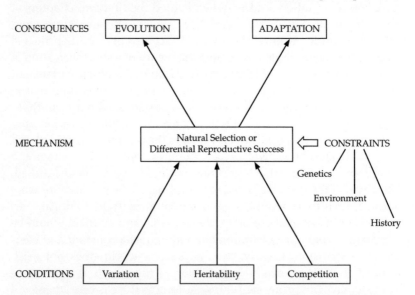

The components of modern evolutionary theory. Natural selection, or differential reproductive success, lies at the heart of the theory. Other components act either as the conditions necessary for natural selection to occur, or as constraints on how it occurs, or as a consequence of it having occurred.

upon certain conditions. Evolution is the outcome of those conditions, and evolutionary patterns will vary if those conditions vary. The fact that they do accounts for the enormous diversity of forms of life and the pattern of evolution itself. Some of the conditions – particularly competition – vary more than others. The genetic system, for example, is common to all organisms, and there is a limited amount of variation in how it operates, largely dependent upon the presence or absence of sexual systems of reproduction. Genetic systems in fact are not only conditions necessary for evolution, but also major constraints. The particular pattern of inheritance found in all living systems – the Mendelian system – means that inheritance is confined to parents and their offspring. Under natural conditions it is not possible to pass on genetic information or characteristics to any individuals other than offspring. If such a thing were possible, then evolution might be expected to occur in a radically different way and perhaps much more chaotically. Indeed, it can be argued that this is the case with cultural change, where items of cultural information can be transmitted from any individual to another. As a result what might be referred to as cultural evolution is much more complicated than biological evolution. However, it is primarily the latter that is of concern here.

Apart from evolution – change through time – there is another consequence of natural selection. This is adaptation. In its simplest meaning this refers to goodness of the fit between an organism and its environment. The better fitted an individual is to its environment, then the better adapted it is. Adaptation is a consequence of natural selection because it is those individuals who are better adapted to their environment who will leave more offspring, and given the other conditions, then over time a population will come to be adapted to its environment. Without the underlying principle of natural selection, though, there would be no reason to expect an adapted suite of plants and animals. Adaptation also provides a basis for looking at the features of any animal and asking why they occur at all. We can now see how the grandiose question – why are there human beings? – can be broken into a number of more easily answerable questions – why are humans bipedal? why do they have large brains? and so on. Natural selection and the principle of adaptation permits the dissection of the human being into pieces and allows us to delve into the problem of why humans have been put together the way they have.

I have considered the nature of evolution in some detail. Why? One reason is that, despite, or perhaps even because of, its simplicity, it is a frequently misunderstood theory, and its basic

principles are therefore worth rehearsing. Another is that while evolution is often the ultimate object of investigation, in practice what is
of concern to most evolutionary biologists is the changing pattern of
conditions and constraints. An evolutionary analysis is not a simple
mapping of change through time, but an attempt to place that
change within a framework of the constraints and conditions of the
biological world. That, indeed, is the underlying purpose of this book
– to show how the pattern of human evolution discovered from the
fossil world and preserved in the features of ourselves and other
animals is understandable in terms of the fundamental principles of
evolutionary biology.

Humans before Humanity

Whatever the mechanism, the most easily grasped notion of evolution is that there is change though time. This allows us to envisage
(and the fossil record can document this) a world in which there
were no humans. However, prior to the appearance of humans,
given the continuity of the evolutionary process, there must have
been something similar to humans, and before them something yet
again a little more different, and so on back into the primeval soup.
Continuity between all living things is an essential element of evolutionary thinking, and remains one of the most far-reaching
consequences of the Darwinian idea.

The continuity of living forms is a powerful fact from which much
can be inferred. The first and obvious one is that all living matter
shares certain characteristics. The most universal trait is DNA
itself, the replicating molecule of life, and this provides evidence for
the fact of evolution. The more closely related any organisms are, the
greater the number of characters they share. Continuity provides
the evidence on which reconstructions of the history of life on the
planet and classifications of all living things are based. Continuity
thus underlies the principle of comparison in biology, and it is these
comparisons that throw light on how and why evolution took and
takes place.

Our focus is not with the whole of evolution, but one very small
portion of it. Humans are just one species in a family that includes up
to twenty species, in an order that contains twenty or more living
families and many more extinct ones. And that order (the primates)
is just one of over twenty-five orders of mammal. We could continue
up through the diversity of life to the estimated billion species on the
planet today – not to mention the even more numerous extinct ones.
And it is the extinct ones that matter here. A direct consequence of

the acceptance of evolutionary ideas has been the active search for extinct animals and for past phases of life on Earth. Nowhere has this been more active than in human evolution. There may be only one living species in the hominid family, but in the past there have been many more.

The extinct hominids demonstrate the continuity between humans and other apes. In other words, they show that the principle of continuity applies as much to humans as to other animals. Extinct hominid species also show that the perceived gap between humans and other animals – one basis for arguing that evolutionary ideas do not apply to humans – is an illusion. It is created by the accident of extinction. If living apes become extinct, then the gap will become greater; if a living Neanderthal were to be found in the tundra of Siberia, the gap would have become smaller, and yet nothing would actually have changed in humans. Fossil hominids also show that the world has in the past been populated by intermediate forms. These intermediate forms are the missing links beloved by journalists when any fossil is found, and by creationists largely because they are missing. They are also the humans before humanity. They provide the direct evidence for evolution having occurred and for the path it has taken. They give us a basis for testing and refuting specific theories of human evolution. They are, too, the proper comparative basis for considering humans as they are today. And, most importantly here, they provide evidence for the specific context in which human evolution occurred, and for the details – the time, the place, local rather than global – that actually determines the path of evolution. Humans before humanity were not truly humans, but species in their own right that survived in many cases for hundreds of thousands of years. They did not exist because they were 'evolving into humans' but because they had adaptations that enhanced their survival. They are the legacy of the Darwinian revolution that can best answer the question of why there are humans at all. Without Darwin's ideas most of them would not have been found, or if they had, they would have been beyond scientific interpretation. The 'humans before humanity' of the title ('hominids' in the scientific terminology used here) that have been discovered in the fossil record provide the direct evidence for our past. Their significance has historically lain simply in their existence, as evidence for the evolutionary process itself. My intention here is to try to link the pattern of the humans before humanity to the conditions that determined their evolutionary fate. Humanity itself is but one part of the story.

During the course of this book I will examine these other types of creature, and will try to reconstruct their lives and their behaviours,

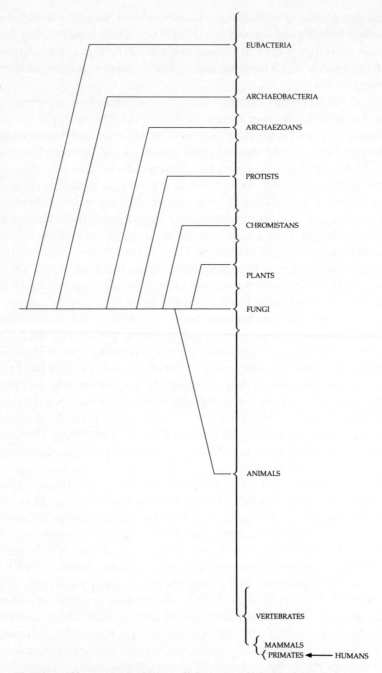

EUBACTERIA

ARCHAEOBACTERIA

ARCHAEZOANS

PROTISTS

CHROMISTANS

PLANTS

FUNGI

ANIMALS

VERTEBRATES

MAMMALS
PRIMATES ◄——— HUMANS

The place of humanity in the overall diversity of life. Humans are just one species, in one order of mammals, in the phylum of the vertebrates, which as part of the animal kingdom belong to just one of eight of the basic types of life on the planet.

to map their appearances and disappearances. There is little doubt that they are among the most interesting animals that have ever lived. They must have lain somewhere upon a continuum between humans and the other primates, and yet they became extinct. They can therefore illustrate the evolution of complex animals in complex environments. However, as they are also in some cases our ancestors, in other cases close evolutionary cousins that did not survive, they are also informative about the path to humanity. What will emerge is that these long gone creatures, whose worlds can only be dimly perceived, are necessary to reconstruct the evolutionary biology of our own species, because they are the only way to pinpoint the time and place, the context, in which being a human made evolutionary sense. It would be a mistake to look upon these hominids either as humans, or as stepping stones on the way to humanity, but none the less they can indicate why there are humans now.

To put these creatures into context it will be necessary to look at when they lived and their relationship to living humans. But first, viewing humans as biological organisms rather than divinely created heroes poses a challenge to science.

3

What are Human Beings?

Apes and Angels

At the height of the debate about evolution Benjamin Disraeli was asked whether he thought man was an ape or an angel. He replied with apparent conviction, 'Now I am on the side of the angels.' Most people are less certain. While many religious fundamentalists and a few die-hard Darwinians would mirror Disraeli's certainty, for most people the question remains a great imponderable. The compassion inspired by the latent humanity that lurks in the eyes of a gorilla is testimony to the close affinities humans have with the apes. And yet the distance can be all too easily amplified by comparing the crude vocal utterances of a chimpanzee with the poetry of Shakespeare. No ape could aspire to the technology of the computer world, or to the complex social organization of even a small human community with its web of kinship and friendship or political machinations.

This, though, is the paradoxical position in which many find themselves. Humans are descended from something like an ape, and yet they are significantly, perhaps irrevocably different. Somehow most people manage to maintain these two, apparently mutually exclusive, views simultaneously.

How the human mind manages this may be yet further evidence of the extent to which humans have developed a subtle and powerful brain. That brain is one of the symbols of the difference between humans and the rest of the animal kingdom. It is several times larger than it should be, and is clearly essential to human survival and evolutionary success. The modern human brain is approximately 1400 grams in weight. In general, brain weight is closely related to the overall body weight of an animal, and were humans to have the size of brain expected for an animal of our size, it would be about 500 grams.

That large brain undoubtedly facilitates a wide range of distinctively human abilities. Most obvious is linguistic skill. Humans can utter a vast array of sounds, but more importantly, they can order them using diverse and flexible rules, and they can understand the meanings, overt and implicit, of those sounds. Humans can create images in any number of media, and, again, give order to them. They may create order in the material world as well, with technologies that may be brilliant in either their sheer simplicity – the boomerang – or their complexity – spacecraft capable of collecting data from distant planets and sending it back to Earth.

None of this would have occurred if all human activity were based on individual prowess. A million monkeys typing for a million years might produce the works of Shakespeare, but they would be unlikely to be able to manufacture, market and distribute those typewriters in the first place. That would require considerable co-operative behaviour and an organized society. Human society is based on complex networks of social and economic interaction – co-operation, competition, dependence, altruism, friendship and enmity.

All this is an aeon, almost literally, from the world of the African apes that are our closest living relatives. While one may accept that the fossil record shows the slow putting together of the bits and pieces that make up the physical human being, it takes a defiant mind to believe that the behaviour that goes with it has been cobbled together in the same random and haphazard way. Humans are descended from apes, but have made themselves angels. Most people learn to live with this apparent paradox – siding with the apes against the angel of death of the fundamentalists, while at the same time taking the part of the angel when faced with the threat of the animal within us.

Apes and angels may be symbols in the general course of human ideas over the one hundred years since the development of the Darwinian framework, but they also represent the several strategies that scientists, both social and biological, have used to pursue the question of the nature of being human in a post-Darwinian world. Ironically the symbols may have survived, but their meaning has been transformed radically over the years.

To Disraeli and his contemporaries there could be little doubt about the nature of either apes or angels, although few people could have seen either. Humans and angels were made in God's image, and even when the former fell below these high ideals, there was little question about their potential nature. This spiritual potential could be buttressed by the ideas, prominent since the Enlightenment, of both the perfectibility of man and the improvement in the conditions in which he might live. As many authorities have pointed out, the positive side of Darwinism lay in its concordance with the notion of progress. While Darwin himself emphasized the mechanism by which evolution operated – natural selection – his contemporaries read much more importance into the linear story that this mechanism produced. Indeed, a great deal of the work carried out in the fifty years after Darwin's death was concerned with maintaining the concept of evolution while rejecting or tinkering with the mechanism.[1]

Against this social, political and religious background, animals were, in contrast, a blacker, darker concept. Tennyson's 'nature red in tooth and claw' was a powerful image, reflecting both the Darwinian ideas of competition and the general view of the jungle as a world without rules. Despite some magnificent observations of natural history, the day-to-day lives of animals were by and large unknown, particularly those, such as apes and monkeys, living in remote parts of the world. In their place were tales of the murderous behaviour of gorillas, the child-eating habits of baboons, and the raucous and promiscuous lusts of the monkeys. The contrast between this and the ideal of even a fallen angel was so great that there could be little room for even a close relationship between the behaviour of apes and humans, let alone a simple continuum between them.

Alfred Wallace, Darwin's co-founder of the theory of natural selection, saw this problem as overwhelming.[2] To him natural selection was simply not powerful enough to bridge the gap in behaviour between animals and humans. His solution was to formalize this gap, and to propose that while natural selection produced the basic diversity of life, and led to adaptation among plants and animals, the brain of humans was the product of divine intervention.

This solution is one that is echoed in much later work on the evolution of human behaviour, and lies at the heart of the recent debate on sociobiology, where the tenets and principles of behavioural ecology were accepted for animals, but their application to humans was both questioned and rejected. Humans were a unique and special case, beyond the power of normal evolutionary processes.

Darwin himself was less certain of this, and explicitly extended his theory to incorporate humans. In general terms this was done in *The Descent of Man*,[3] which is principally a catalogue of similarities between humans and other animals, particularly primates, running the whole gamut of characteristics from the digestive tract to morality. However, Darwin recognized that the principal problem in applying evolutionary principles to humans lay in the realm of human behaviour. In a classic example of his talent for tackling major theoretical issues through meticulous experimental design he developed two strategies for dealing with these difficult problems. The first, seen in *The Expression of the Emotions in Man and Animals*,[4] was to concentrate on the interaction between behaviour and anatomy. While this might appear trite in comparison with the big questions about incest and warfare, this work elegantly demonstrated the central point – that behaviour could evolve. Because the muscles of the face were so delicately related to the ability to make expressions, and those expressions were in turn related to feelings and emotions, and therefore situated in a social context, then in principle all elements of behaviour could be subject to natural selection and could therefore evolve. While the example was trivial, the conclusions were extremely far-ranging and formed the basis for much that has followed in ethology and behavioural ecology.

Darwin's second strategy can be found in his *Selection in Relation to Sex*. Natural selection works through differential rates of reproduction for different individuals in the context of competition for scarce resources. Darwin's theory of sexual selection extends this principle. Because individuals, particularly males, compete for females, then that competition, independent of any competition for other resources, will result in the operation of a selective mechanism – namely, differential rates of reproduction. Because of the direct relationship between competition for mates and reproduction, sexual selection has the potential to occur very rapidly – often in the runaway form proposed by Fisher.[5] Where one sex is choosing mates on the basis of some characteristic – say the colour of a bird's plumage – then selection can happen very rapidly, as in each generation the most extreme form will be selected, as will the preference for it in the other sex. These two will thus fuel each other, and lead to extremely rapid evolution.

Darwin's discussion of sexual selection was an important advance in the understanding of the evolution of behaviour. Wallace's objection to the application of selective mechanisms to human behaviour was based on the fact that many characteristics found in both humans and animals did not seem to affect their chances of survival and ability to acquire resources, avoid predation and so on. However, by focusing on reproduction, the struggle for sex rather than the struggle for survival, Darwin was able to show that competition, and hence the potential for selective mechanisms, occurred

Gorillas, the largest of the living apes, were thought to be aggressive, violent and dangerous, as is shown in this Victorian print. In fact they have been shown to be very shy and seldom pose a threat to any human.

under all circumstances, and applied to behaviour in the most general sense. Furthermore, because in many species reproduction occurs in a social context, then behaviour cannot be considered in a narrow context, but must extend to social forms as well.

It turns out that the co-founders of modern evolutionary theory had established the framework for much later debate about the nature and evolution of human behaviour. On the one hand, Darwin focused on the selective mechanisms by which behaviour could be treated within the same framework as morphological characteristics such as the shape of bones or the size of muscles, and was therefore a true evolutionary problem. Furthermore, behaviour in humans was on a continuum with that of other animals. Wallace, on the other hand, claimed that selection was too weak a process, and adaptation too much an agent of fine tuning, to be extended to all realms of organic diversity, and in particular, to complex and sophisticated behaviour.

Much twentieth-century work on the evolution of behaviour, human and animal, played out in departments of zoology, psychology and anthropology, in the laboratory and in the field, has been a continuation of this debate. The status and standing of apes, angels and human uniqueness have risen and fallen as new results and ideas have emerged.

The retreat of the angels has to a large extent been a response to changes in knowledge about both humans and animals. While sightings of angels have not increased markedly over the last hundred years, apes have become considerably better known. The development of long-term field studies in which individuals within any primate population can be recognized has shown that animal behaviour is far from the playing out of instinctive stereotypic routines. Primates in particular live rich social lives, develop strong attachments to particular individuals, vary their behaviour according to context and intent, and operate within what can only be termed a social structure. Furthermore, such behaviour is mediated through complex neurological processes that indicate considerable overlap between the cognitive skills of non-human primates and those of humans. It is not that ethologists have shown animals to be complex beings living in some sort of harmonic idyll, for there is also plenty of evidence for aggression, competition, and even behaviours that result in the death of individuals. Apes and monkeys are not the amoral beasts of Victorian legend, nor the innocents of a pre-human Utopia implied by more romantic writers. Instead, they are complex, variable, flexible and often intelligent in ways that can be both altruistic and devious.

From the point of view of human behaviour this discovery has led to both new insights and new problems. Apes, and chimpanzees in particular, have been shown to possess characteristics of behaviour and cognitive skills approaching those known in humans. Wild chimpanzees make and use tools, are capable of sophisticated communication, can plan and execute courses of action over long periods of time, and manipulate both objects and other individuals to their own ends. Each of these abilities observed among chimpanzees has shown apes crossing the various rubicons that have been set down as marking the boundaries between human and non-human. For many years, for example, man was referred to as 'the tool-maker', unique among animals in his ability to shape objects to his needs. Jane Goodall's,[6] Adrian Kortlandt's[7] and subsequent observations[8] have shown that this is not the case. The same has been shown for meat-eating, symbol manipulation and language, all at one time thought to be markers of humanity. In terms of apes and angels, man might be an angel, but it is not at all clear that the apes might not be as well.

Along with the growth of knowledge of animal behaviour has come a greater understanding of the diversity of human life and, to some extent, a loss of confidence in the extent to which humans could be said to be on a pedestal above the swamp of animal brutishness. The camps of Dachau and Belsen, the millions killed in religious wars, the extent of poverty, famine and disease, and the almost boundless capacity of humans to do damage to each other at national and personal levels have, in the twentieth century, rather dented human self-esteem. Furthermore, as social anthropologists have revealed the richness and complexity of so-called primitive life, and to some extent implied that the simpler the society the greater the harmony and level of individual happiness, the more difficulty people have had holding on to the Victorian notion of a ladder of progress climbing closer and closer, hand in hand with technological and economic development, to the level of the angels.

It seems that the apes have become more angelic during the course of the twentieth century, the angels, or at least their human representatives, more apish. Where it was originally thought that humans were the advanced and progressive form of life (the angels), and other animals the more primitive, now it may be argued that the animal within us is our noble side, and humanity or civilization the blacker side – a complete reversal of the original Victorian image.

Defining Humans

Ironically, evolutionary ideas have been one of the principal casualties of this change in perspective. Darwinian ideas gained wide acceptance during the late nineteenth and early twentieth centuries, but these were not necessarily the central ideas that Darwin himself wrote about. As noted in chapter 2, evolution, to Darwin, was the outcome of natural selection and adaptation, and therefore was a by-product or consequence of the biological mechanisms in which he was most interested. It was these consequential ideas of change and progress that caught the public and scientific imagination, and indeed the mechanism was itself rejected or diluted by many. However, without the mechanism of natural selection evolutionary biology can be little more than a descriptive enterprise, and the concept of evolution becomes prey to those who see it as synonymous with progress. From the idea of progress it is only a short step to a ladder of complexity, a classification of advanced and primitive forms, and an evolutionary framework that relates primarily to value judgements rather than to scientific objectivity. Against the background of an increasing scepticism, particularly in the social and human sciences, about the progressive nature of human culture and behaviour, it is hardly surprising that evolution became an idea that had outlived its usefulness in human affairs.

One of the central problems is to decide how to characterize the human species and human features. Clearly, to explain why they should occur, we need to know what they are. The conceptual divide between apes and angels has acted as a barrier to coming to any such understanding. Apes and angels are really just ideals, and extremely nebulous ones at that. If evolution is a dynamic, shifting process, these static ideals are inadequate, and force both social and biological scientists into harder and more extreme positions, with a major obstacle to communication between them arising as a result. At one extreme, humans might be seen as nothing more than naked apes, and we can cheerfully and uncritically throw every evolutionary method at them in the hope of unravelling their mysteries. At the other, we may decide that in the process of becoming human a rubicon of evolution has been crossed that itself washes clean our evolutionary ancestry.

The proposed scientific and evolutionary perspective clearly prevents the adoption of the second course of action. Although there may have been subtle shifts in the way in which selection may operate among humans, it is unlikely that everything biological has been transformed. In seeking an evolutionary understanding of humans the tactic of ignoring everything about humans that is difficult to encompass biologically, and settling instead for the safer territory of anatomy and physiology, must be avoided. Instead humans should be defined in an all-embracing manner, and then the adequacy of Darwinian evolution as an explanation can be tested. Such a definition itself becomes both interesting and elusive once the relative security of the present species is abandoned for the unknown of the fossil past – the humans that lived before humanity.

Living humans are characterized by a number of universal features. These underpin the fact that despite some superficial differences both within and between populations, all humans belong to the same species – that is, they are all capable of interbreeding and producing viable offspring under normal conditions. They also attest to a common evolutionary origin. Many of these features are simply observable in terms of anatomy. Humans are bipedal, that is they walk in an upright manner on two rather than four limbs. Bipedalism is probably the most physically obvious human feature. It is unique among primates, and is an adaptation that has had a pronounced effect on the entire musculo-skeletal system. To accommodate this form of locomotion the lower limb has been elongated and strengthened, the foot arched and buttressed and the grasping ability found in other primates lost. The pelvis has become shortened, rounded and flared to act as a primary support for the upper body. The vertebral column has become curved and strengthened in the lumbar region. The head is set vertically upon the vertebral column, with the hole through which the spinal cord passes to the brain being relocated at the base of the skull instead of towards the back. Although it might have been possible to be bipedal without changes to the forelimb, this has become shortened and is less mobile at the shoulder joint.[9]

Bipedalism may have made possible a number of other human features, in particular manual dexterity. Humans, along with other primates, have grasping, sensitive, five-fingered hands. Most primates are capable of high levels of manipulative dexterity, but this is found in its most extreme form among humans, who have thumbs that are capable of opposing virtually any of the other digits.

Other anatomical features are also striking. Humans have very large brains for their body size, and faces that have become reduced and flat. The hair over most of our bodies is miniaturized (not absent, as is implied by the concept of the 'naked ape'), and the skin therefore exposed. Humans also have a potential for copious sweating.

The other distinctive characteristics of human structure lie in the reproductive organs and secondary sexual characteristics. Although modern humans are only moderately sexually dimorphic in terms of size – on average females are about 84 per cent the size of males – there are quite marked secondary characteristics. Females have, in the dry terminology of the comparative anatomists, pendulous breasts, and more rounded and fleshy buttocks. Physiologically they generally have large fat deposits that they can draw upon during periods of nutritional stress. Men are generally more hirsute, although this varies in extent from population to population. It is perhaps also reassuring to know that the male human penis is large compared with the other apes, although the testicles are, compared with a chimpanzee at least, not especially large.[10]

These are the relatively obvious features that all humans share in

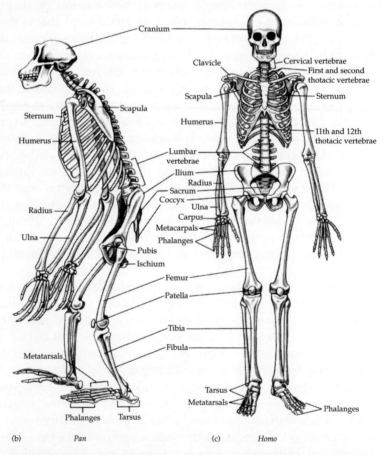

The skeletons of humans and chimpanzees show the characteristics of shared ancestry, and the differences that make each of them unique. The human skeleton has been greatly modified by the evolution of upright, bipedal walking.

their physical make-up, but they tell little about what these features make possible and are used for. Later it will be important to find out why they exist at all, but here all that is sought is a superficial view of what humans have achieved in an evolutionary sense. Most would agree that it is not so much anatomy as human behaviour and mental ability that are the real marks of the species.

A whole suite of behavioural characteristics can of course be found in humans and a claim made for their uniqueness. Not surprisingly many of these have been selected as *the* feature that made humans the way they are. Man the tool-maker, man the hunter, woman the gatherer, *Homo economicus*, *Homo hierarchicus*, *Homo politicus*, and *Homo loquans*, these are all sobriquets that have been designed to epitomize human nature. They, and several others, are all traits that have been used by various people to identify the driving force that underlies human nature.

To many, tool-making has been the decisive factor. Even a cursory glance at the world shows that humans depend on technology to an extraordinary extent. This is not just the case for urban, industrial peoples, but for all societies. Houses, food, weapons, games, all involve technology to some extent even if they are relatively simple in construction. It is hardly surprising, especially when humans are compared with other animals, that it has been suggested that this is the key trigger to human success. The basis for tool-making stems partly from the manipulative skills of dexterous hands and partly from the ability of the brain to co-ordinate and create actions that have technological consequences. The practical applications of this are obvious, from the simplicity of the wheel to the power of the nuclear reactor. The significance, though, is broader than just the tools themselves. What technology does is to allow humans to modify and create the world they live in. Technology can make a species the active component in the construction of the environment, in contrast to the fate of most species, that are generally seen as the passive recipients of the world into which they are born. The best they can do is to react to their environment in ways that maximize their chances of survival. If humans want a predator-free environment then they build a house into which predators cannot come. If they want a warm environment, a fire can be lit. Technology is the means by which the human world is created.

Although technology has lost some of its appeal as a marker of humanity as its less desirable consequences have become apparent (pollution, nuclear arms and so on), none the less this remains an important element of what it means to be human. However, that humans alone are tool-users and tool-makers is no longer

acceptable. From the lowly termite to the hammer-using chimpanzee, other animals clearly use tools too.[11] More important
perhaps, other species cannot be viewed as simply accepting their
environmental lot. They are not just passive receivers, but like
humans are actively affecting their habitats, food resources and
shelters. If this is the case, then technology alone cannot be the
trigger that set humanity on its course. What other characteristics
might be important?

However ineptly, all humans use technology in their daily lives.
The same cannot be said for another feature which has attracted the
attention of scientists concerned with human origins. Hunting as a
means of survival is not pursued extensively today. Rather, it is confined to a few groups of indigenous foragers such as the Kalahari
San, the Eskimos, and the Australian aborigines. Ten thousand years
ago it would have been far more useful. Today, although agricultur-
alists will supplement their diet with hunted food where possible, by
and large most animal food comes from domestic sources. Hunting,
though, whatever the particular context, may have been an important element in human evolution.

The logic of this statement runs as follows. A look at the anthropoid primates shows them to be primarily vegetarian. The diet will
vary from species to species and even within species, but plant foods
represent the primary food source of all of them. This can be contrasted with humans, who are by no means carnivore, but most of
whom eat meat at least some of the time, and also tend to value it
highly. The disparity between humans and their close evolutionary
relatives led Raymond Dart, most notably, to suggest that this was a
key to becoming human.[12] Why should hunting have been so important? Partly this derives from the idea that animals are harder to
acquire than plants. A plant just sits there waiting to be eaten. An
animal on the other hand will run away or even fight back. Eating
meat is often thought to be the reserve of either very strong animals
or very clever ones. For small, defenceless bipedal humans it was
intelligence that was required. More than this, hunting seemed to
require co-operation between individuals (and therefore social
organization) and language to co-ordinate activities. Early humans
were not just hunters, but social hunters, and so hunting implies
more than just the eating of meat.

Robert Ardrey, in his book *The Territorial Imperative*,[13] added
another element to the hunting hypothesis. He associated the act of
killing animals for food with the psychological state of aggression.
Hunting therefore acted not just as a behavioural trigger but also as
a psychological one – the killer ape – a theory which had a strong

appeal in a century in which killing of humans by humans has been almost a normal action.

Hunting is certainly an interesting behaviour, not least because it is something that may have been more important in the past than it is today. However, the hunting hypothesis has lost support in recent years. The rise of vegetarianism and a concern for both conservation and animal welfare may be one reason. More significant, perhaps, has been the influence of direct observations of living hunter-gatherers by anthropologists such as Richard Lee among the Dobe !Kung of Botswana, which showed that in practice meat was not such an important part of the diet (about 20 per cent), and that gathering of plant foods was the primary subsistence base.[14] If this is the case for modern hunter-gatherers, then how much more so is it likely to be the case with ancient ones. If modern humans either could not or would not hunt vast quantities of meat, then why should our ancestors have done so? Furthermore, studies of wild carnivores showed that social ones, such as lions or hyaenas, were quite capable of hunting without sophisticated communication. Equally, evidence has accumulated that chimpanzees can and do hunt.[15] Hunting no longer seems the appropriate trigger for human characteristics.

Language has also been suggested as the unique feature that makes possible the human species.[16] Only humans have either the extraordinarily wide range of sounds and meanings or the associated grammatical structures that allow a multitude of meanings to be built up. Language is itself closely tied to social behaviour – humans communicate so prolifically because they live in complex social groups, and are bound to each other by webs of kinship and friendship, and levels of interdependence that are unknown outside the human species. This inter-dependence is particularly strong in economics, and it has been suggested that both the sexual division of labour (males and females have different economic spheres) and the more complex economic roles that arise with specialization are essential platforms on which human society, and hence humanity, has been built. The main point here perhaps is that as humans are, above all, social beings, then what humans are is closely tied up not with humans as individuals but with humans as part of humanity as a whole.

If we cannot disentangle language from social and economic behaviour we may be lead on to the basic notion that what makes us human is culture. Anthropologists use the concept of culture in myriad ways, but the heart of them is the idea of a cognitive template upon which the whole structure of human behaviour is shaped. Its pivotal element is that it provides the flexibility to enable all sorts of behaviours, thoughts and actions to be modified and for widely disparate activities to be integrated. Man the culture-bearing animal can replace and embrace all aspects of humanity, from technology to politics to aesthetics.

The Evolutionary Catch 22

As I sifted the evidence for defining humanity, there was a shift from anatomy to behaviour and from behaviour to cognition, the intellectual landscape became increasingly slippery. The difficulty in coming to any clear conclusions and the inability to put the finger on a single characteristic represents the tactical problem of investigating human evolution, and of unravelling the ape from the angel. In many of the arguments there is in fact a Catch 22 that is liable to unsettle simple discourse.

One such catch is the way in which people use arguments about apes and humans. On the one hand, we can compare the two groups, or indeed humans with any other animal, and note the ways in which humans are unique. When identified, be they tools or language or whatever, these can be erected as the key feature underlying human nature and uniqueness. The catch comes though when, having identified these as uniquely human features, the inference is drawn that there are no parallels in the natural world, and so it is no good looking to other animals for their evolutionary source. An evolutionary approach is therefore brought to a dead end, and we are left with angels, pure and simple. The catch is neater than this, though, for it has usually been the case that when a trait has been identified as special to humanity, chimpanzees have been found to possess the same trait. No sooner was Kenneth Oakley's *Man the Toolmaker*[17] published than Jane Goodall reported chimpanzees using tools.[18] When hunting was popular, then chimpanzees and baboons were found to hunt.[19] When humanness withdrew to language, all the studies of language acquisition in apes were produced.[20] The catch comes full circle when it is then pointed out that as these features occur in other species, they cannot be what made us human. A further knot is sometimes added, as in the case of human hunting, when it can also be shown that humans do not really do what they are supposed to do anyway. This circle of evolutionary fallibility can be repeated for virtually any trait, and it serves to prevent the breakdown of the ape and angel ideals that have dominated the imagery of human evolution.

Another Catch 22 can be found in the difference between single-factor models and multiple-factor models. A single-factor model is one where a single trait is identified – say language – and then everything is seen to derive from this one feature as a series of dominoes falling over under a simple cause-and-effect process. Language leads to sociality, sociality leads to co-operation, co-operation leads to hunting, hunting leads to tool-making, tool-making leads to greater intelligence, and so on. Usually these models fail to satisfy because they break down when it is realized that the sequence could make as much sense the other way round. The early models tended to be single-factor ones, and they were replaced in the 1960s and 1970s by ones recognizing the complexity of feedback between parts of the system. Language leads to sociality, but sociality leads to language. A systemic, feedback model, showing the inter-relatedness of everything, is much more likely to reflect reality. The catch comes in, however, in that when the models are built, showing that everything is related to everything else, we are left little the wiser. Not much is known about the causality involved, and the image of early humans becomes little more than a reflection of modern ones. What remains is the catch that either a model fails because it is too simple or else it fails because it is so complicated that it is meaningless.[21]

The final catch is that knowledge is constrained by the way that humans are today. They are the only guide, and yet they are also the end point of the very event that we are trying to unravel. Humans are the consequences of the evolutionary process, and we are in the position of having to work out the causes from those consequences. The problem is, though, that the traits that are so important today – language, technology, kinship and so on – may not have been the same as the ones that made the species successful in the first place. Computers are successful today because they are word processors, games, means of communication, but their origins and development were related far more to their ability to crunch numbers and make large calculations. The danger lies in making the same error with human evolution, and imposing as causes what are in fact the consequences, and so failing to recognize that what is trivial now may have been central in earlier times and vice versa. Human features may not be adaptations to some past environment, but *exaptations*, in Gould and Vrba's terminology – accidental by-products of history, functionally disconnected from their origins.

The way to break out of these catches is also the way to break down the extremes of apes and angels. Part of the solution is to recognize that the apes and the angels are but the starting and end points of the journey, but that in between lies something that is completely different. Furthermore it is necessary to remember that the path between the two is not a straight one, but rather, a jagged and broken one, with many branches that lead nowhere. The key to tracing that path is to give primacy to the ape, not because humans necessarily are apes, but because the only thing known for certain is that our ancestors once were apes, and to find out if and when humans became something else, then it is best to assume that the evidence will fit well against that ape-like template. Only then will it be known when and how it differs. More importantly, perhaps, it is necessary to depart from the generalities of how human origins might be approached and set out to answer some of those small questions that in the previous chapter seemed the way forward.

4

When did we Become Human?

The Problem of Time

A recent survey in the United States aimed at testing the scientific knowledge of the population yielded the slightly alarming result that over 60 per cent of people thought that humans were contemporaneous with the dinosaurs. No doubt this was at least partly the result of films such as *2 Million Years BC*, which showed Raquel Welch as a cave-girl among the heavy-footed dinosaurs, but it also reveals a general lack of understanding about the temporal dimension of history. While there may be little point in knowing the exact evolutionary chronology any more than knowing that the Battle of Hastings was in 1066 rather than 1067, the overall scale is important. If the human species is only 2000 years old this has very different implications for a number of questions than a date of 2 million years (close to the right answer) or 200 million years (closer to the dinosaurs). Part of the problem no doubt lies in the difficulty that anyone has in even comprehending vast periods of time. When the Battle of Hastings, at less than 1000 years ago, seems totally remote, how much more difficult is it to understand a period of over a million years.

Human conception of time enables us to comprehend times that mean something personal – the number of days before a holiday or the time someone might deserve in prison for stealing a car – but beyond these human lengths the whole operation becomes rather haphazard. Humans' own longevity is such that most people can happily cope with the time periods which they or those around them have witnessed. In practice this means grandparents – three generations, or a period of about 75 years. To that extent, for someone in their thirties or forties now, this means the First World War represents about the limit of time that can be at least indirectly comprehended personally. For people in their twenties it will now be the Second World War that forms this baseline. When we think just about recorded history – from the beginning of the Christian Era, for example – this period will include some eighty generations. To go back to the origins of the human species involves having at least a passing comprehension of the idea of 8000 generations. Obviously we neither can nor would want to know what happened in each one of those generations, but it is necessary to understand that in this sort of context the events of a single, or even a dozen or a hundred generations is insignificant. The same humanly-ordained units and concepts cannot be applied to a period of time that is beyond the type of analyses and interpretations that help to order and explain recent events. It is the role of evolutionary biology to show how the long and the short term can in some way be united, and provide a meaningful answer to the question, 'when did we become human?'

This question is significant for reasons other than simply being able to say 'a long time ago' or even 'a very long time ago'. Philosophically there may be implications for understanding the effects humans have had on the environment or the stability of forms of human behaviour and adaptation, while technically there may be important implications in terms of genetic diversity. Assumptions are often made about how long people have lived in certain areas, and therefore about their rights to land or resources. The significance of, for example, the 40,000 years that Australian aborigines have lived in Australia is dependent not only upon the relative length of time of European colonization (about 200 years) but also on the length of time that humans have been present elsewhere in the world.

Of course, it is a relatively simple thing to ask the question 'when did we become human?', and it is tempting to give a very trite answer. However, it turns out that this question is actually quite difficult to answer. There are two reasons for this. One is that the fossil record, the primary source of information about the distant past, is a partial record, and only a censored image of that past is possible from the few remains that have been preserved. There is always the chance that a more remote fossil will be found that will take back human origins another million years or so. This is something that has happened repeatedly throughout the history of research on human evolution, from the discovery of the first australopithecines in South Africa in 1924 (thought to be about one million years old at the time) to the discovery of 'Lucy', or *Australopithecus afarensis*, in the 1970s (about 3 million years) or the 4 million year-old *Australopithecus ramidus* found in 1994.[1] A million years here or there may not matter a great deal in some ways, but when this constitutes a doubling of the antiquity of the human lineage there may be significant implications.

The other reason is that while it may be possible to put fairly precise dates on the origins of particular groups of hominids and of our own particular species, it is not always apparent what this may mean in terms of the origins of humanity or of humanness. Finding a fossil that has teeth that are similar to ours back at five million years ago may be important, but it does not necessarily follow that this creature was a human. Indeed, the whole point of this book is to show that being a human and being a hominid are by no means the same thing. To ask the question 'when did we become human?' virtually traps one into the answer 'it depends upon what you mean by human'. And it is all too clear that different people will use different criteria to establish whether or not a particular population is or is not over the line into humanity.

The Long, the Short and the Bipedal

What is necessary is to look at some of these criteria and see how they provide different answers to the question of the antiquity of humanity. Two problems influence the way this question might be answered: the first is whether the criteria leave evidence in the fossil record; and secondly, are they useful? Are they really informative about humans, or are they just incidental markers of a continuous, long-term, and gradual evolutionary process? It will soon emerge that this question is more complex than appears at first sight.

This complexity has long been acknowledged. Over the 150 years since human evolution has been recognized as a serious problem there have in fact been two very different perspectives on this issue. They may be termed, for simplicity, the 'long' and the 'short'. On the one hand, there is a tradition of seeing humans as having a unique ancestry stretching far back into the remote past. While the exact length of time has varied as our perception and ability to measure the geological past has changed, it has been believed that humans are sufficiently distinct from other animals that they must have evolved independently for a long period of time. This perspective has in turn been fuelled by the quite natural desire of palaeontologists to find earlier and earlier evidence of human fossils – after all, no one would win the Nobel prize for finding the second oldest fossil.

On the other hand there are those who have been impressed by the similarity of humans and other animals, especially the primates and the anthropoid apes. Where the long perspective sees differences, the short perspective is overwhelmed by the similarity. The inference to be drawn is that humans diverged only relatively recently from other animals, and therefore have only a short independent evolutionary history.

The battle between the long and the short is one that has characterized the history of human palaeontology. We can trace the history of both of these, and the contest between them and the debate that has occurred since the publication of the first book on human evolution – Darwin's *The Descent of Man*.[2] Darwin himself was essentially a long-chronology person. Although he had no fossil record and only a limited grasp of the extent of geological time, none the less he believed that human antiquity was great. His reasons for believing this are interesting.[3] To understand them it is necessary to remember that at that time the chronology recognized and accepted by most educated people in Europe was excessively short – that is, the chronology based on biblical history. This suggested that the world was only about 6000 years old. A major scientific controversy of the nineteenth century was the attack on the Bible as a source of geological information. Lyell, Darwin's great mentor and the founder of modern geology, was the man largely responsible for establishing the idea that geological time was larger by several orders of magnitude. His concept of uniformitarianism – that is, that only processes and mechanisms observable in the present should be used to explain events in the past – was one that emphasized the need for much longer lengths of time for geological (and biological) events to happen. Under the orthodoxy of the Bible, time was in very short supply – according to Genesis the world was created in seven days. In order to explain all the historical events that could be observed – the building up of mountains, the massive deposition of sediments, erosion of whole landscapes – in a very short period of time, it was necessary to evoke processes involving vast amounts of energy. This was the sort of energy that could lift up a mountain in a matter of hours, or create the enormous transgressions of the sea that could be observed in extensive marine sediments found high up on those mountains and that were 'witnessed' in the great biblical flood. In other words, a short chronology demanded that geologists were extremely economical with time, but spendthrift with energy.[4]

It was against this background that Darwin wrote his theory of evolution. His main problem was to persuade people that natural selection was an appropriate mechanism for change in the biological world. To do that, he used Lyell's principle of uniformitarianism and

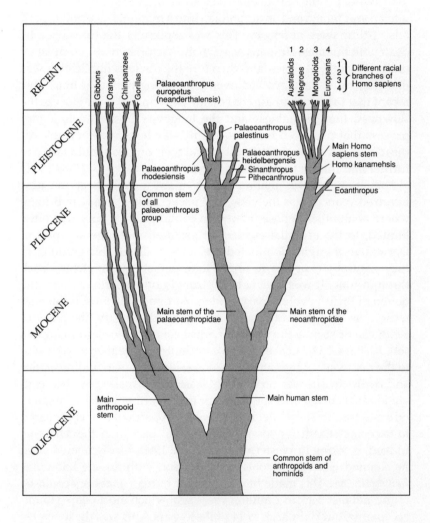

An early reconstruction of the evolutionary relationships of humans, apes and fossil hominids, by Sir Arthur Keith. It shows a very ancient divergence between apes and humans, and deep evolutionary divisions within living populations and for fossil groups. Compare this with the modern estimates shown in the figure on p. 58.

invoked only mechanisms that he could observe in the world around him. Consequently, given the type of mechanisms he did observe – changes from generation to generation through variation in natural and artificial breeding – he recognized that the change must be very, very slow, and therefore gradual. This meant that for him to be able to convince people that an amoeba could evolve into a fish and a fish into an amphibian and an amphibian into a mammal, enormous periods of time were necessary. This was especially the case when he dealt with humans. Humans were, to the Victorian mind, completely unlike any other species, and therefore the obvious implication was that they must have evolved over a very long period of time. This meant that to make his case convincing Darwin had to emphasize the slowness, the gradualness and the longevity of the evolutionary process that produced humans. In this way he was responsible for the idea that the search for humans and their origins should take one further and further back into the fossil record.

This tradition was maintained by many people over the next hundred years. It was the view held particularly strongly in Britain, where anatomists, whose interest in evolutionary theory was often limited, to the essential element of gradualism as the key part of Darwinian thought, dominated the subject. Given this gradualism they naturally expected to find a long perspective stretching back through time. It was a view held strongly by Sir Arthur Keith, the doyen of British evolutionary studies. As one of the most influential writers on human evolution in the twentieth century, the effect of Keith can be seen in the work of Louis Leakey and indeed in Louis's son, Richard.[5] The Leakeys have contributed more than anyone else to the discovery of human fossils, and their success in finding earlier and earlier fossils was no doubt at least partly fuelled by their conviction that such fossils should occur much further back in time than others dared to speculate. Louis Leakey in particular was prepared to take the search for fossil humans well back into the Miocene. Indeed, working in Western Kenya in the 1960s he discovered what he claimed were stone tools in association with an ape known as *Kenyapithecus*. This made him argue that human ancestors could be found not just back to 2 million years ago, which most people found acceptable, but right back to 14 million years. This was the long perspective with a vengeance. The tradition was continued throughout the 1960s and 1970s with such controversies as the hominid status of *Ramapithecus*.[6] These debates over specific fossils were merely a continuation of this ancient debate.

It must be stressed that it was not just the inherited beliefs of Darwin that produced such support and interest in the long perspective. Clearly the evidence of anthropology, psychology and neurobiology, all of which seemed to stress differences between humans and other animals, appeared to underpin the idea that humans were radically different. It was a combination of this radical difference in structure and behaviour, in association with the belief that evolution must be gradual, that essentially provided the key to a widely held view that humans had taken a long time to evolve and were distinct from 'ordinary' animals far back in the past. That such a view was also more acceptable philosophically

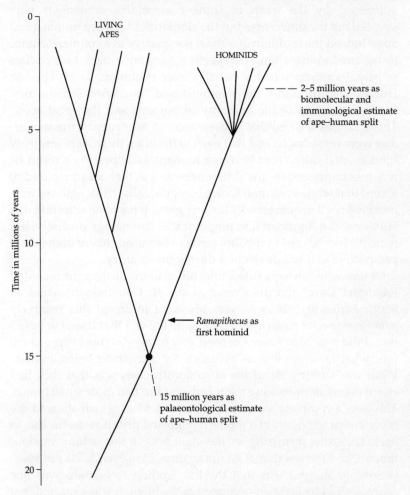

The estimated timescale of hominid evolution at the time of the *Ramapithecus* debate in the 1960s and 1970s. If *Ramapithecus* was a hominid it indicated a split between apes and hominids occurring around 15 million years ago. The first molecular and biochemical estimates for the age of this event suggested a date of between 5 and 2 million years ago.

and in terms of compatibility with religious beliefs may well also have been a significant factor in the predominance of the long perspective.

However, although the long perspective has probably been the dominant one for the last hundred years it has not held the field entirely. Indeed, virtually all the controversies in the history of human fossil discoveries, from Dart's *Australopithecus africanus* to Richard Leakey's '1470', have been fuelled by the counter-belief that the evidence could not support a long chronology. This evidence was buttressed by the work of some comparative anatomists who stressed not the differences but the similarities between humans and apes. Indeed the tradition of a short perspective as a counter-balance to the predominant long perspective, goes right back to Darwin's principal supporter in his debates over evolution, T. H. Huxley. Huxley was primarily an anatomist and he carried out the first detailed analyses of the anatomy of humans and the great apes.[7] Those studies showed that in basic anatomical structure the similarities were remarkable, and that even in the brain there were few truly fundamental differences between humans and apes. As a result he was less impressed by the differences, and in turn was prepared to accept that while evolution took place gradually, the lengths of time involved need not necessarily be very great. It has to be admitted that Huxley was not prepared to pinpoint that chronology precisely, but none the less he could probably claim to be the originator of the short perspective as it has developed during this century.

Others who perhaps fitted into this pattern of thought included Raymond Dart,[8] the discoverer of the first australopithecines in South Africa in 1924, who was prepared to accept this relatively primitive species as an ancestor despite the fact that it was younger than Piltdown Man (later exposed as a forgery[9]). This brings to the fore what is acceptable as evidence for something being human. What was striking about the australopithecines was that they had relatively modern-looking teeth and jaws but had quite small brains. This was a complete contrast to Piltdown Man, which showed the reverse characteristics of a modern brain and primitive teeth. This in itself is hardly surprising as the skull was in fact a fully modern human, and the jaw that of an orang utan. However, what Piltdown seemed to suggest was that the line leading to humans was one characterized by early enlargement of the brain. It was expected that this, the key distinctive characteristic of humans, was what must have evolved first. However, if the australopithecines were true ancestors of humans, then large brains must have evolved relatively late, and by implication fairly rapidly. As such, Dart's discovery was rejected, or at least taken very critically, partly on the grounds that it was against the expectations of the long chronology.

In more recent times the prominence of the short-chronology approach has become more and more important. That this has come about has been largely the responsibility of the primatologist Sherwood Washburn.[10] He made two significant contributions. The first was to encourage the study of primate behaviour and ecology. He argued that living primates displayed characteristics that would not have been inappropriate for the early hominids. He showed that they were living in complex social groups, with well defined individual roles and relationships, capable of flexible and adaptable responses. This suggested that the differences between modern humans and living monkeys and apes was not as great as had been thought, and therefore did not require a great deal of time to evolve. A close relationship with other primates implied a short chronology.

His second great contribution was that he stimulated one of the first detailed attempts to try to specify when humans and great apes diverged, and indeed to establish who were our closest relatives. Washburn's approach was to encourage research in biochemistry and molecular biology, and it is this approach, more than any other subject, that has led to the triumph, in the last few years, of the shorter chronology for human evolution.

This then is the background for considering the question 'when did we become human?' On the one hand it can be argued that the differences between humans and the rest of the biological world are large and a long period of time is involved. On the other, recognizing chimpanzees as very close relatives means that it is necessary to seek only a short period of time since the evolutionary divergence from the apes. The only way to resolve these questions is to examine the details of the fossil record and the evidence of comparative biology. However, in doing so, what becomes apparent is that the reading of the fossil record is to some extent dependent upon how the significance of the observed features is interpreted. What this contrast between the long and the short perspective shows is that there must be some expectations about the scale of human evolution dependent upon observations of humans today, and secondly, that there must be criteria for testing whether those expectations are fulfilled in the fossil record. However, while there is noticeable continuity and even a changeless element in the debates of palaeoanthropology, there is progress as well. This is perhaps most marked in the overall acceptance of the basic chronology of hominid evolution and the recognition that there are different problems within the overall pattern – the origin of the hominids is different from the origin of *Homo*, which is in turn different from the origins of our own species. It is this greater resolution that allows us to focus on the significance of time and place in human evolution, and to go beyond the ancient debates.

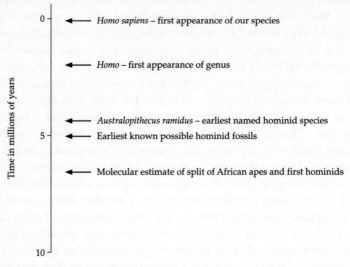

The scale of human evolution based on current evidence. Compared with Keith's scheme shown on p. 53, this has a much shorter overall timescale. Nothing in hominid evolution can be earlier than the late Miocene, and differences within *Homo sapiens* are small and young.

The Path to Humanity

Against this background of conflicting ideas about the antiquity of humans, perhaps the best approach is to look at the evidence for when we did become human. This is primarily, at the outset at least, a question about evolutionary relationships and evolutionary rates. The classic measure of evolution is referred to as phylogeny – the reconstruction of evolutionary relationships between groups, species and higher taxa. What this amounts to is trying to work out the sequence of separations or branching events in evolution, for the evolutionary process is itself one of divergence of populations from ancestral common ancestors. Having worked out the branching sequences, the next task is to assign dates to the various events. The pattern and processes of evolution can then be reconstructed from this phylogeny. In terms of human evolution the branching sequence can be used to pinpoint when humans had their origins. These principles can be used to explore not just the question of when we became human, but also the scale of humanity in the context of life as a whole.

The earth is estimated to be about 4.6 billion years old. Life (usually defined as chemical systems capable of replicating themselves) did not appear until 3.5 billion years ago, and multicellular plants and animals evolved about 750 million years ago. Vertebrates made their appearance 450 million years ago, and from then they colonized both land and sea. The early dinosaurs emerged about 200 million years ago and dominated the earth until their rapid disappearance about 65 million years ago. Although the mammals have their origins at least 150 million years ago, they did not spread and diversify until after the extinction of the dinosaurs, at the same time as the radiation of the flowering plants or angiosperms. The primates, the biological order to which humans belong, only evolved during the last 60 million years or so. Mammals, primates and humans belong only to the most recent evolutionary events, and in the largest perspective are mere newcomers, arriving only in the last few minutes and seconds of the evolutionary clock. Put another way, 70 million years ago none of the living primate species existed; by 10 million years ago probably about 50 per cent of the 180 species were in existence, and

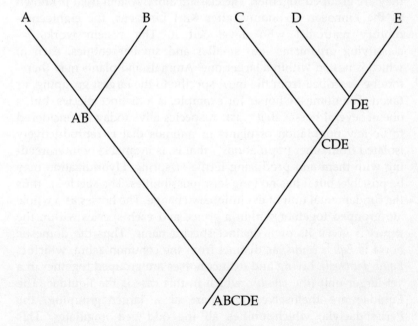

Evolution can be thought of as a series of branching events, as lineages diverge and begin to evolve independently. The living species (A,B,C,D, and E on the diagram) are subsequently united to each other (their common ancestors – AB, DE, CDE, etc.) through time and more ancient evolutionary histories. Phylogenies are essentially reconstructed in the reverse order in which they occurred in evolution.

by 1 million years ago over 90 per cent had evolved. Among mammals the average life expectancy of a species is about a million years,[11] and so it should be expected that most of the species living today, including humans, have appeared only very recently. However, there is considerable variation, and while the general pattern of evolution can provide some expectations of when humans evolved, it is only by looking at the details of the apes and monkeys, and the fossil record itself, that this can be pinpointed more exactly.

The simplest approach to answering the question 'when did we become human?' is to ask the even simpler phylogenetic question – when did human ancestors diverge from our closest living relatives? It is a very attractive notion to argue that the first humans were those alive when a lineage began to have its own independent evolution and was no longer part of a common ancestral stem with the apes. This of course depends upon being able to say which is our closest relative. This is not quite the simple task it may seem.

For biologists to communicate with each other it is necessary for them to agree about what particular animals are called and how they are grouped together. The classificatory system used is known as the Linnaean taxonomy, after Karl Linnaeus, the eighteenth-century naturalist who developed it. The system works by classifying organisms into smaller and smaller entities, each of which is nested within a larger one. Animals and plants may therefore be described from the most specific to the largest grouping, or taxon. The domestic horse, for example, is a distinct species, but is one of several horses that exist. A species alive today is considered to be any population of plants or animals that is reproductively isolated from other populations – that is, is incapable of interbreeding with them and producing fertile offspring. Hybridization may be possible, but it has no long term possibilities. The species is thus the fundamental unit of evolutionary change. The horses as a whole are grouped together within a *genus*, and each species within the genus is given its own distinct species name. Thus the domestic horse is *Equus equus*, as distinct from the common zebra, which is *Equus burchelli*. Living and extinct horses are grouped together in a yet larger unit (the *family*), which in this case is the Equidae. The Equidae are themselves just part of a larger grouping, the Perissodactyla, which unifies all the odd-toed ungulates. This process can continue upwards to include all the mammals, all the vertebrates, all multicellular animals and ultimately all the living world. The Linnaean taxonomic system provides a shorthand for describing animals and their relationships. If we want to refer to all horses we can use the word, 'equid', whereas to refer specifically to

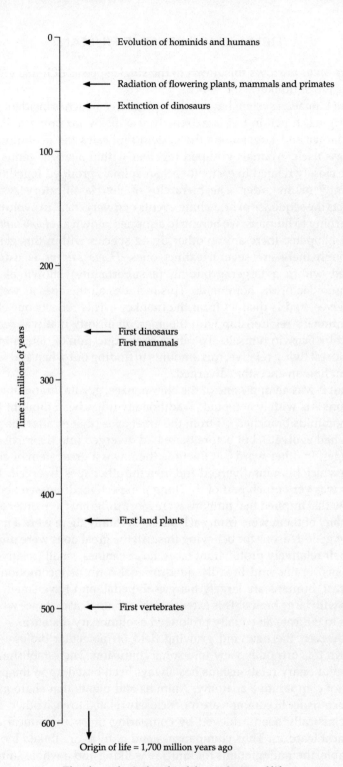

The chronological scale of the evolution of life.

Equus equus narrows this down to the single species of horse today, exclusive of donkeys and zebra.

The Linnaean system has another advantage which is incidental to its original function but is extremely useful. By grouping animals into larger and larger units the system replicates the evolutionary process itself. Animals grouped together within a single genus are more closely related to each other than animals grouped together in a family or an order. The hierarchy of the classificatory system reflects the sequence of branching events or divergences in evolution.

Turning to humans, we belong to a species known as *Homo sapiens*. As it happens, there are no other living species within this genus, although there are several extinct ones. *Homo sapiens* is usually placed within a larger grouping (a superfamily) known as the Hominoidea or the hominoids. This includes all the apes as well as ourselves, and is distinct from the monkeys. This reflects our close evolutionary relationship with the apes, an affinity that was recognized by Darwin himself. To return to the question of determining the closest living relative, this amounts to finding out when and from whom human ancestors diverged.

That it was an ape – one of the chimpanzee, gorilla, orang utan or gibbons – is widely accepted. Traditionally it has been thought that the hominids branched off from the great apes (that is, after the gibbons had evolved), but before these had diverged into their existing lineages. In other words, at the time there was a great stem of apes, from which humans diverged and then the other apes diverged. This view was very much part of the 'long perspective' discussed above, a view that implied that humans were more different from other apes than any of them were from each other. Certainly there were a number of good reasons for believing this. All the great apes were united by their relatively protuberant faces, large canines, small (relative to humans) brains and broadly quadrupedal form of locomotion. In contrast, humans are largely hairless, bipedal, and have small, flat faces with large brains. This extensive morphological distance would seem to indicate an equally prolonged evolutionary distance.

However, the new and growing field of molecular biology has thrown this orthodox view into some confusion. The establishment of evolutionary relationships has always been based upon the principle of comparative anatomy. Animals and plants that share more characteristics in common are more closely related to each other. This has classically been achieved by comparing gross anatomical and physical features. Thus chimpanzees and gorillas are linked by, for example, their adaptations for knuckle walking and a whole suite of other skeletal traits. This principle works at any level and applies to

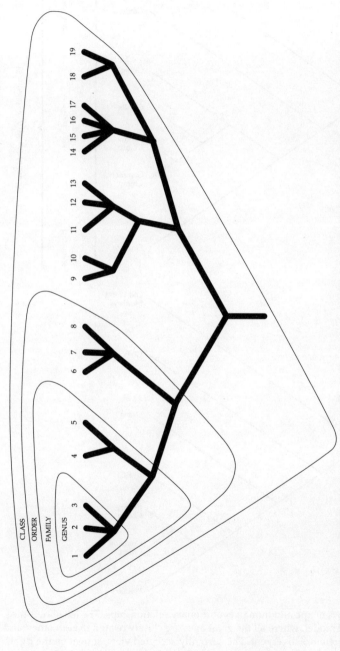

The system used to classify organisms is known as the Linnaean system. The nineteen species shown here (1, 2,...19) are classified into larger and larger entities or categories. In the hypothetical example shown here, these are *genus*, *family*, *order* and *class*. The term hominid refers to a family in which there are two genera, *Homo* and *Australopithecus*. The family Hominidae includes all the species that have evolved since the split from the chimpanzees, and which are more closely related to humans than chimpanzees.

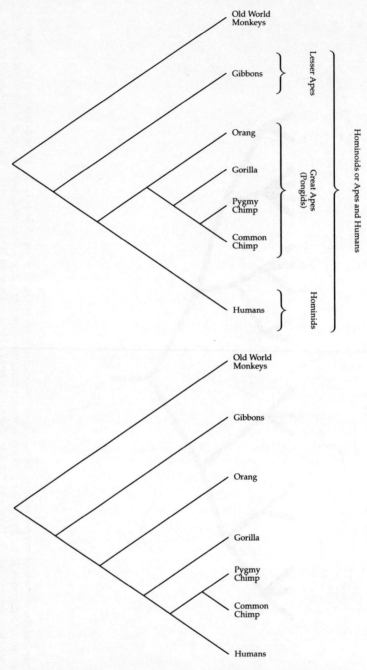

Two views of ape and human evolutionary relationships. The left shows the traditional model, where all the great apes are closely related to each other, and humans are more divergent. The currently accepted view, shown on the right, shows that chimpanzees are more closely related to humans than they are to the other great apes.

biochemical and genetic levels of organization as well. As the structures of proteins, amino acids and ultimately of DNA itself have been decoded, it has been possible to use the morphology of the molecules themselves to classify species and to reconstruct their evolutionary relationships. Thus, for example, two species that share the same sequence of amino acids for a protein are more likely to be related to each other than they are to a species that has a different one, and so on right down to the actual sequence of bases in a string of DNA.

This is, of course, hardly surprising as the separation of organisms in evolution is a process of genetic divergence – it is genetic difference that leads to the reproductive isolation that underlies the definition of species to start with. What is perhaps surprising is that morphology at the gross level and genetic structure at the molecular level do not always show the same results. The application of molecular techniques to a wide range of organisms has frequently yielded a situation where species that are morphologically distinct have an almost identical sequence of genes, while species that are very similar at the anatomical level are genetically divergent.[12]

It has been learnt that the process of evolution is a lot more complicated than the incremental addition of odd bumps and ridges on skulls and skeletons. The phylogeny of the hominoids is a case in point. Genetically the great apes proved not to be as closely related as their superficial anatomy would suggest. Whether proteins, amino acids, immunological distance or actual gene sequences are compared the results are the same. The major divide among the great apes and humans lies not between humans and the great apes, but between humans and African apes in relation to the Asian great apes, the orang utan. Chimpanzees, gorillas and humans are more closely related to each other than any of them are to the orang utan. The great apes are not what is termed a true or natural branch or clade in evolution. This means that rather than humans diverging from the apes before they themselves had split, the gibbons and orang at least had already undergone independent evolution. Humans are just another type of African ape.[13]

This discovery has become well established. It suggests that in the context of hominoid evolution as a whole, humans are a relatively recent rather than a very ancient lineage. However, there is an even more startling possibility. When the same molecular techniques are applied to humans and to the African apes it is very difficult to determine the branching sequence among them. Intuitively what might be expected is that humans diverged first and then the chimps and gorillas from each other. Most of the evidence, however, indicates that it is not possible to discern the sequence precisely.[14] Where it is possible to determine the order of divergence it seems that gorillas separated first from the common ancestor of chimpanzees and humans, and only subsequently did humans and chimpanzees diverge.[15]

Humans, therefore, are specifically African apes and seem to be closely related to chimpanzees, far more closely in fact than anyone had previously thought possible. Only a relatively small number of genes separate these two species, despite the enormous number of morphological differences. The question of which is our closest relative has been answered. While the answer is the chimpanzee, which is not surprising, it is important to note that it is specifically the chimpanzee and not the great apes as a whole. This is an important distinction that will be returned to in later chapters. What has not been answered is the question of *when* this divergence took place. Certainly it is not as ancient a separation as one might think, given that so many other evolutionary events – the divergence of gorillas, orangs and gibbons, for example – had already taken place.

Obviously the fossil record can be examined to see whether it is possible to calibrate these events. This, though, would be premature, as the potential of molecular biology as a source of evolutionary information is not yet exhausted. Genetic differences between species do contain the seed of knowledge about rates of evolutionary change.

Evolutionary change can be thought of in two ways. One is, as Darwin himself thought, the change that results from animals adapting to their environments as a result of natural selection. The rates of change would therefore not be constant, but would vary according to the intensity of the competition or the amount of environmental change, or any number of other factors. However, evolutionary change, as it is actually measured, is nothing more than the rate of genetic change, whatever the cause of those changes might be. Certainly if the rate of genetic change is a haphazard response to environmental changes, with periods of very fast change and very slow change alternating, then genetic distance alone cannot give information about the date of evolutionary events. If, though, the rate of genetic change is constant then the quantity of genetic distance between any two or more species should indicate not just the sequence of events (their relative position in time) but also their exact timing (an absolute position in time).

Is there any basis for assuming or expecting that the rate of genetic change is constant? That there may, at some levels, be constant change is the basis for the molecular clock.[16] The amount of DNA in any single organism, be it an amoeba or a human, is vast. In the case of humans there are some 30,000,000 bases, or some 100,000 genes. This is far more than there is any need for from the point of view of developmental biology. It seems that there is a vast amount of DNA which has no known function – often known as junk DNA. This has the characteristic of having no observable effect on the phenotype as it develops. As such it is immune from the effects of selection. If it changes it does so not because of changing fitness and selection in response to adaptive needs and the environment, but purely independently. Such change as there is occurs through the process of mutation – the 'random' occurrence of errors in the process of gene replication. This is the way in which new genetic material, the raw material of evolution, appears. Mutations by and large appear at a constant rate, and if the products of this mutation have no selective consequences, then the accumulation of new genes, the accumulation of genetic differences, will be an accurate clock measuring the rate of evolutionary change.

When this principle is applied to the hominoids the results are intriguing. The first living great ape to diverge was the orang utan, and the molecular estimates for this event are about 12 million years ago. It has already been established that it is virtually impossible to distinguish clearly the exact sequence of events amongst the African apes and humans, but it appears that these three lineages separated somewhere between 6 million and 8 million years ago, towards the end of the Miocene. If chimpanzees and humans diverged later than gorillas, then this is more likely to have occurred at the younger end of this time range.

Humans, then, have been in existence for about 6 million years. This might provide one possible answer to the question, for that is the length of time that there has been an independent evolution separate from chimpanzees. How does this fit in with the expectations for a long or short chronology? By anyone's standards 6 million years is quite a long period of time. However, by the standards of much of the debate, this is in fact a remarkably short period of time. When the molecular results were first appearing most palaeontologists took the view that hominids diverged from (all) the apes at least 15 million years ago, and according to some authorities a figure of 30 million years was probable.[17] Furthermore, all the events in anthropoid evolution were assigned dates of very great antiquity, and they have all shifted downwards. At the time, a fossil known as *Ramapithecus punjabicus* from the Siwaliks in Pakistan, dated to at least 10 million years ago, was thought to be an early hominid, and provided support for the long perspective. The molecular dates cast doubt on this interpretation, doubts that were confirmed when more complete specimens were discovered showing that in fact *Ramapithecus* was more likely to be an ancestor of the orangs than of humans. (Indeed, under the endless reclassification of fossils that occurs, the unfortunate *Ramapithecus* has not only lost its status as the first hominid but has now disappeared as a valid taxon altogether. It is not only the Stalinists who remove all trace of their past mistakes.)

Overall, the findings of this new approach seem to indicate that humans are a young lineage, not an ancient one. From a geological perspective it is only recently that humans had an independent evolution. This conclusion, though, must be qualified in a number of ways. The first is that while the molecular clocks appear to have considerable scientific validity, they are not quite as simple as has been made out.[18] As more and more genes have been examined it is clear that there is some variation in the rate of genetic change and that neither the assumption of constant mutations across the genome nor the assumption of adaptive neutrality hold entirely. What this means is that there is not a single clock, but several clocks, ticking at different rates, some slower, some faster, some keeping a more constant time than others. While the results of molecular biology provide robust indicators of the timing of evolutionary events, they do not necessarily give an absolutely accurate answer. It may therefore be profitable to look at the findings of the more traditional source of information about the evolutionary past, the fossil record.

Molecular biology provides a series of expectations about human origins. Assuming the methods are accurate, then there is no point in looking for the first humans amongst the Jurassic dinosaurs, despite the wishful thinking of Hollywood producers. Instead there is the prediction that the earliest fossils showing the path to humanity should occur in the Late Miocene – that is, between 10 and 5 million years ago, allowing for the possibility of inaccuracies in the molecular estimates. It is possible to be more precise still, not so much in terms of time, but in terms of geography. The phylogeny of ape genetics indicates that humans and the African apes are the closest relatives. Chimpanzees and gorillas are in fact relatively restricted in their distribution, confined to the central parts of Africa, stretching east and west, but not north or south. As Darwin himself pointed out: 'In each great region of the world the living mammals are closely related to the extinct species of the same region. It is therefore probable that Africa was formerly inhabited by extinct apes closely allied to the gorilla and chimpanzee; and as these two species are now man's nearest allies, it is somewhat more probable that our early progenitors lived on the African continent than elsewhere.'[19] Molecular biology points the same way that traditional comparative

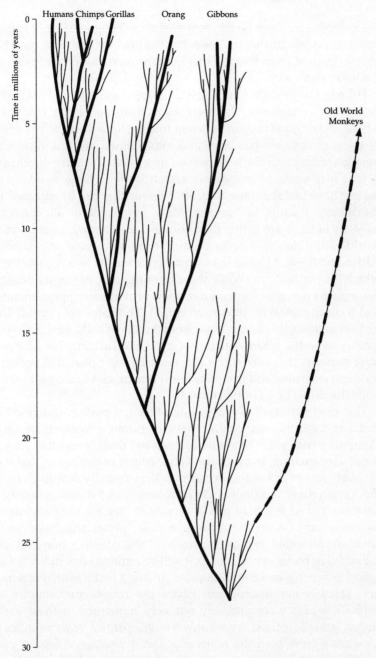

A molecular chronology for human evolution. The thick lines show the living
species and their direct ancestors. The lighter lines show the lineages that
evolved at various times, but which have become extinct. Molecular biology
can inform us about the history of living groups only. Palaeobiology can
provide information about the other groups, namely those that provide the
evolutionary context in which hominids evolved.

anatomy did for Darwin. His next sentence, though, is that 'it is use-less to speculate further on this subject'. However, it is now possible to test the prediction that the first humans were in Africa about 6 or 7 million years ago.

Human fossils were found first in Europe and then in South-east Asia. After a number of false trails, it was really only in the 1960s that Africa, the original favourite, began to come into its own. The story of these discoveries has been told many times, and the historical details are not particularly important here.[20] What is more significant is that fifty years of intensive research by very large numbers of people have yielded more than 3000 fossils that can be assigned to the lineage leading to humans. Most significant of all, the vast majority of these are earlier than human fossils anywhere else in the world. The limestone caves of the Transvaal region of southern Africa and the Rift Valley in eastern Africa are the two regions from which the fossils come. While these localities may not be the sites of the actual origins of humankind, they provide an approximation, and a confirmation of an African origin. The dates are crucial. The earliest indisputable fossil comes from the site of Lothagam, a remote locality in northern Kenya to the west of Lake Turkana. The lake and river deposits that surround Lake Turkana have provided perhaps the most extensive and complete fossil record, and Lothagam repre-sents the earliest part of this record.

The specimen itself is fairly insignificant, a portion of mandible that has teeth that are characteristic of humans rather than apes. Compared with some of the more dramatic finds from other sites it is not very exciting, but its significance comes from the fact that it is probably just over 5 million years old. Very recently new finds from the Awash River in Ethiopia, dated to around 4.5 million years ago, have confirmed not only this approximate age for the first line to humans, but also, in the form of a new species (*Australopithecus ramidus*), provided the most primitive and ape-like hominid yet found.[21] The molecular evidence, it will be remembered, indicated an age of between 6 and 8 million years, and so a first hominid at 5 mil-lion years is not unexpected. Given the remote probabilities of animals, which were probably not very numerous in these early stages, being fossilized, a date only a million or two years younger is reassuring from both the molecular and the palaeontological per-spective. There are in fact some earlier fossils, dated to around 6 or 7 million years ago, from an area to the south of Lake Turkana, but these are even more fragmentary and cannot be assigned with any certainty to the human or ape lineages.[22]

After 5 million years ago the fossil evidence for humans picks up considerably. Sites in Ethiopia dated to over 4 million years old, and a site in central Kenya, Tabarin, at just under 5 million years, are all supportive of the existence by then of a human lineage. From 3 million years ago the fossils are both more common and more complete. Where Lothagam and Tabarin provide only fragmentary evidence for human characteristics through the dentition, the material from Hadar in Ethiopia is far more convincing. In particular, a single specimen, known colloquially as Lucy and more formally as AL-288, found by Don Johanson in 1974, consists not just of dental and cranial fragments, but a nearly complete skeleton.[23] Head, teeth, arms, back, hips and legs are all present. What this specimen shows is that by 3 million years ago there was in existence an animal that was for all intents and purposes bipedal. Upright walking is perhaps the most distinctive characteristic of humans, and of Lucy, with her elongated lower limbs, flared and rounded pelvis, and distinctive angle to the head of the femur. While there are clear differences when compared with a fully modern human, such as relatively long arms and curved phalanges, there is little doubt that this is part of the human lineage. By 3 million years ago, hominids were more bipedal than any other ape, clear evidence perhaps that by then something had become human, if bipedalism can be accepted as a key characteristic. Indeed, bipedalism itself was probably older. Apart from the fossils themselves there is also evidence from footprints which are 3.7 million years old. At a site in Tanzania called Laetoli, Mary Leakey discovered a number of footprints which had been made in wet ash after a volcanic eruption.[24] Along with footprints of all sorts of typical savanna animals was a sequence of imprints which were distinctly human – no evidence for four limbs being used and a characteristic stride pattern with the heel and the ball of the foot clearly pronounced. Here was direct evidence, even if it is not possible to say exactly who made them, of an upright, bipedal species.

It can be argued, perhaps, that the answer to the question 'when did we become human?' is as follows: about 6 or 7 million years ago on the basis of the molecular evidence for when the human lineage departed company from the other African apes, or about 5 million years for sure, on the basis of the first palaeontological evidence, and certainly by about 4 million years ago, when clear evidence exists for that unique human feature, bipedalism. Does this add up to humanity?

From Bipedal Apes to *Homo sapiens*

The problem has so far been considered in terms of two criteria: phylogeny and overall anatomical structure. By one interpretation of phylogeny it can be argued that a distinct human clade can be identified as far back as 5 or more million years ago. In terms of overall anatomical structure, if bipedalism sets humans apart from other species more than any other characteristic, then humanity stretches back more than 3 million years. However, both these criteria are open to alternative interpretations and there are other criteria that need to be considered.

Phylogeny is primarily concerned with determining branching or divergence events in evolution. Only one such event has so far been established, the branching of a line leading to humans from the one leading to the other African apes. This implies that this event was, if not the only event, at least the most significant. However, this is far from the case. The fossil record shows evidence from at least 3 million years down to less than 50,000 years ago of many types of human fossil. Dating and morphology suggest that these do not simply belong to a single evolving lineage, but rather, constitute a number of distinct evolutionary trajectories. The conclusion that must be drawn is that far from being a simple linear progression, the line leading to humans is itself made up of a number of branching events. Between 3 and 1 million years ago there were at least two distinct lines of human evolution – one leading to some megadontic specialists known as the robust australopithecines, and another leading to larger-brained forms, our own genus, *Homo*. Even later than 1 million years ago there may have been separate lines of development, one in Europe and Africa, one in Asia; and the neanderthals, a specifically European group, may be a separate line of development compared with anatomically modern humans. Given the probability that there have been a number of branching events, leading to distinct species in human evolution, not all of which could have led to modern humans, then the focus on the earliest appearance of the lineage as a whole may not be the most appropriate one. Perhaps it is the appearance of the genus *Homo* around 2 million years ago that is more critical, or even the appearance of our own species, *Homo sapiens*, that is the crucial event.

To determine this it is necessary to look at the anatomical patterns, for they can give a better idea of the extent to which these fossil groups were more or less like modern humans. If these different types of fossil human are only superficially different from ourselves, then the later branching events may have no real evolutionary

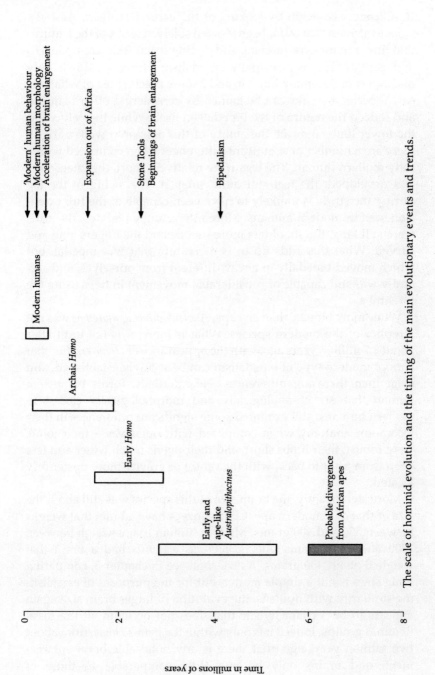

The scale of hominid evolution and the timing of the main evolutionary events and trends.

significance. To begin by looking at the earliest of these, *Australo-pithecus afarensis* has so far been viewed solely in terms of the features that link it to modern humans and distinguish it from chimpanzees and gorillas. This is principally bipedalism. However, this is only one aspect of becoming human, and a more precise idea of what this early species was like can be gained by considering other features, and indeed the nature of the bipedalism itself. While the pelvis and the lower limbs indicate the ability of this animal to walk upright, there are a number of significant differences when compared with a fully modern human. The legs were relatively short, the knee joints less developed, the feet still quite splayed. The swing of the leg during the stride is unlikely to have been capable of the full exten-sion seen in modern humans. More interestingly perhaps, the arms were still long, the shoulders more mobile and the fingers long and curved. What this adds up to is a creature that was bipedal, but which moved bipedally in a way different from ourselves, and sec-ondly was still capable of considerable movement in trees using the forelimbs.

While more bipedal than any ape, *Australopithecus afarensis* was not a replica of the modern species. What is more, it is not until after about 1.7 million years ago, with the appearance of *Homo erectus*, that a more modern type of bipedalism can be securely established, and even then there are differences. Neanderthals, forms of ancient human that are chronologically and morphologically closest to modern humans, still exhibited some significant differences in their locomotor anatomy when compared with ourselves – their joints were robust, their limbs short, and their pelvis much wider and less deep from front to back, with the centre of gravity more posteriorly located.

More significantly, the brain size of this species was still about the size of that of a modern ape. Chimpanzees have a brain that weighs between 350 and 400 grams. Modern human brains weigh between 1300 and 1400 grams. *Australopithecus afarensis* had a brain that weighed about 400 grams. As we shall see in chapter 8, comparing brain sizes is not a simple matter, but for the purposes of establish-ing similarity with humans, the evolution of larger brain size again appears to be a characteristic that does not occur in all the fossil hominid groups. Indeed it is only within the genus *Homo*, from about two million years ago, that there is any noticeable brain enlarge-ment, and brains only become truly comparable to those of modern humans relatively later – less than half a million years ago. For most of the five million years in which human fossils are known, their brains, while progressively larger than those of chimps, none the less cannot be said to be like those of *Homo sapiens*.[25]

Anatomy would seem to suggest, therefore, that while the basic upright body plan of humans was established as long ago as 3 or more million years, other anatomical systems which are strongly associated with being human do not occur until much later. Certainly on the basis of brain size, the term human should be confined to the genus *Homo*, if not to *Homo sapiens* itself.

Perhaps, though, anatomy is not the appropriate criterion, for after all, it is our behaviour that really distinguishes humans from the rest of the animal world. Technology, perhaps may be the key, for as we saw in chapter 3, this has been strongly associated with becoming human. However, although it is known that chimpanzees are capable of rudimentary tool use and manufacture, there are no tools associated with *Australopithecus afarensis*. Indeed, while the first fossils date to nearly 5 million years, the first known stone tools do not appear until shortly before 2 million years ago.[26] If technology is a characteristic of being human, then the bipedal *Australopithecus afarensis* lacks this particular trait. So, too, do many of the later australopithecine species. Furthermore, even when technology does make its appearance, in the form of simply flaked stone tools, there are still a number of discrepancies. The first technology, known as the Oldowan after the site of Olduvai in Tanzania where it was first described, existed for over a million years. Its successor, the Acheulean, characterized by large bifacially flaked axes, remained stable for about a million years as well. Even the later industries associated with the neanderthals, known as the Mousterian and consisting of systematically flaked cores which had been prepared prior to flaking, continued largely unchanged for well over 100,000 years. In contrast, the stone technology associated with modern humans was never in existence for more than 5 to 10,000 years, and generally was far more ephemeral. Furthermore, the rate of technological change among modern humans, continuing through to the present date, is one of accelerating rapidity.

The spatial pattern mimics the chronological one. Modern human technology changes from one region to another relatively quickly, but the technologies of these archaic creatures were homogeneous across and even between whole continents. The Acheulean, for example, although varying in minor form, is known from Cape Town to Cardiff. Comparing modern humans, therefore, with ancient ones in terms of their technology yields a major contrast. Instead of rapid change, flexibility and local adaptability, there is fixed and almost stereotypic production, albeit of forms that may be both sophisticated and elegant.[27] Furthermore, it is only with anatomically

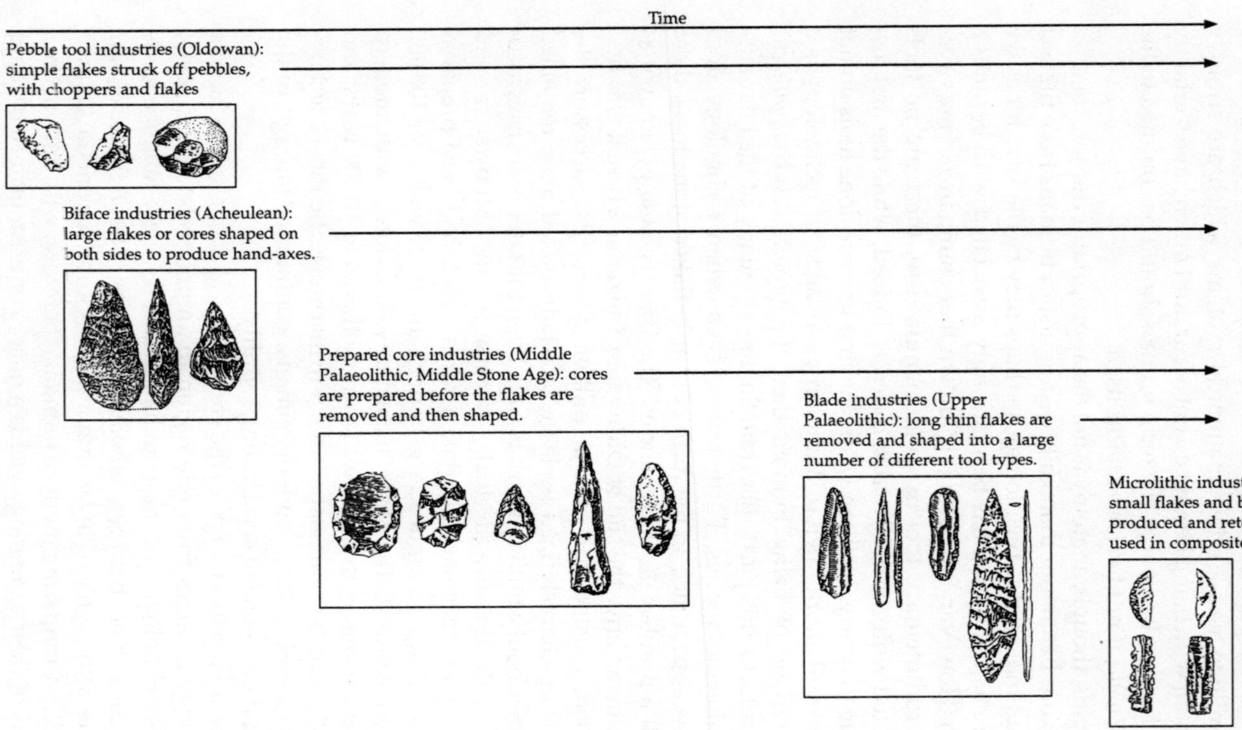

Time

Pebble tool industries (Oldowan):
simple flakes struck off pebbles,
with choppers and flakes

Biface industries (Acheulean):
large flakes or cores shaped on
both sides to produce hand-axes.

Prepared core industries (Middle
Palaeolithic, Middle Stone Age): cores
are prepared before the flakes are
removed and then shaped.

Blade industries (Upper
Palaeolithic): long thin flakes are
removed and shaped into a large
number of different tool types.

Microlithic industries: very
small flakes and blades are
produced and retouched and
used in composite tools.

Technological development in hominid evolution. The pebble tool industries characterize the early genus *Homo*. *Homo erectus* developed the biface tools generally known as hand-axes. Prepared core technologies are associated with archaic and some early modern forms of *Homo sapiens*, while blade technology occurs in some later modern populations.

modern humans, at dates of less than 40,000 years ago, that characteristics such as art make their appearance. Even behaviours such as
hunting would now seem to occur much later in the archaeological
record than was originally thought.

Behaviourally, then, there appears to be a major difference
between modern humans and the rest of the known fossil types.
Behaviour shows the same pattern as the anatomy. Growth patterns,
sexual dimorphism, tooth size, robusticity also seem to show the
same sort of contrast. There is a progressive trend towards modern
humans, but none the less, even the most chronologically adjacent,
the neanderthals, are significantly different.[28] The inevitable conclusion is that the event of diverging from the African apes, the adoption
of upright walking, even the establishment of technology, does not
in itself provide evidence for humans in the sense understood today
– flexible, slowly maturing, lightly built and highly intelligent creatures. Even the overall pattern of the fossil record, showing diverse
trends, seems to undermine the notion of an ancient appearance of
humans. The conclusion that has to be drawn is that becoming
human and being a human being are two different things entirely.
Where does this leave the question – 'when did we become human?'

Apes, Hominids and Humans

So far the term human has been applied in a relatively loose manner.
It has been applied uncritically to anything that lies on the evolutionary path between the divergence from African apes and the
arrival of modern humans. What is now clear, though, is that what
does lie on that path is a very variable group of creatures. That the
path itself is far from straight anyway will be the subject of the next
chapter, but from the chronological perspective that is paramount
here, there are over 5 million years connecting apes and modern
humans, and the connections are not straightforward. As far as can
be told from the fossil evidence, everything along that path (with the
possible exception of the oldest species, *Australopithecus ramidus*) is
bipedal, but not in a form that compares exactly to the modern
human's own form of locomotion. Brain size varies significantly, as
do technology, behavioural flexibility, and growth patterns. In some
cases the links lie more strongly with the apes, in others with modern
humans. Given that evolution is an essentially continuous process of
modification, some trends can be observed, but these tend to be
relatively short-lived and confined to particular groups and lineages.
There are few trends that flow smoothly from the first separation
from the African apes to modern humans. Where among these

trends, though, should the true origins of being human, of humans themselves, be placed?

One solution is to avoid the problem by simply accepting the seamless continuity of evolution. Each point along the path neither is nor is not a human, but is simply part of a process of becoming one. For this continuum it is possible, if necessary, to invent a whole series of terms – missing links, ape men, man apes, proto-humans and so on – to express the inexpressible. It is unlikely that there will be any practical or ethical problems arising from such a classification, so perhaps it does not matter very much. Were a living *Australopithecus afarensis* to be discovered in some remote part of Africa, we might be forced into some rather harder decisions such as whether it should be placed in a zoo or sent to school. There are, though, some other problems with this solution. One is that there is still no answer to the question. Between the first humans at 5 million years ago and the first indisputably anatomically modern humans at about 100,000 years ago is a long period of time in which to bury a problem. More significantly, it is almost certain that most of the known types of fossil human are not directly on the path leading towards humanity in the strict evolutionary sense. Most of them may well be side branches and dead ends, leaving no trace in the modern world. To what extent, therefore, did they contribute to the process of becoming human? Furthermore, the picture of evolution that this seamless view imposes is one of constantly becoming, never of being. Fossils are considered only in terms of what they had been and what they were about to become, never what they actually were at the time they were alive. And yet this is what evolutionary biology is really all about – why an animal is doing what it is, at a particular point in time and space. Evolutionary end points and paths are secondary to this. Therefore to understand when modern humans evolved it is necessary to understand what these ancient forms were like, why they evolved and what became of them; it is necessary to avoid treating them as part of the spectrum from ape to angel. Instead, we must set out to demarcate them and identify each of them in its own evolutionary context. In other words, it is necessary to forego the continuity of evolution, even if temporarily, and instead categorize more strictly the different parts and branches on the path to humanity.

This is a challenge to normal terminology, but fortunately classical zoological taxonomy can provide a sound basis for at least recognizing the problem if not solving it. The Linnaean system described earlier sorts animals into smaller and smaller groups. This principle reinforces the point that the approach to the question of what a human is and when we became human is too coarse-grained. It is not a question of apes and angels, or even apes and humans, but of distinct levels of differentiation. Humans and apes share a level of evolutionary relationship that unites them as members of the Hominoidea – the hominoids. Humans are hominoids in the same way that chimpanzees and gorillas are. This linkage is at the super-family level. The next level down is the family. It is generally accepted that all the species that lie on the path that separates humans from the apes belong to the same family – the Hominidae. *Australopithecus afarensis*, by virtue of its bipedalism and other features, is clearly a hominid, but it is not therefore a human being or human. By definition of its name it is in fact placed in a separate species and genus. This hierarchical separation of the various groups of fossils provides a way out of the dilemma of when humans actually evolved.

We became hominids – distinct from the other apes – when an ancestor diverged from the chimpanzees and gorillas, characterized most probably by an upright posture. This was some time before 5 million years ago. We became human, though, when we achieved the distinctive patterns of anatomical structure and behaviour that can still be found today. This was some time between 150,000 and 100,000 years ago, as it was only then that our own species – *Homo sapiens* – evolved. What lies between, though are a multitude of populations, groups, species that are distinctly hominid – neither apes nor angels. Some of them may be ancestors, most were probably not. However, their importance lies not in whether or not they were ancestors, nor in whether they were truly human, but in the fact that they were 'hominid' – that is, our closest evolutionary relatives. As such they provide the best clues to why and how there evolved a distinct type of hominid – humans. They were the 'humans' before humanity – although the conclusion of this chapter must be that they were not really human. It is these hominids, their distinctive characteristics, their patterns of evolution, their adaptations, behaviours and capac-ities, that provide the central evidence of the process of change from hominoid to human. The fact that they are all extinct is perhaps a dis-advantage, but the task that lies ahead is to unravel the story of these hominids and how they indicate both what lies in the evolutionary space between ourselves and the other living species of animal, and also the reasons why one of them became human.

5

Was Human Evolution Progressive?

The conclusion drawn in the previous chapter was that hominid evolution is too complex to be reduced to a single evolutionary event and a single directional trend. Among the group that can best be referred to as hominids there is considerable diversity in form and behaviour. Furthermore, modern humans represent just the final fraction (so far) of the overall picture. The 'humans' before humanity represent more than just the process of becoming human. What, though, do they represent, and what can they imply about the origins of our own species and the process of evolution itself. This question, and in particular the issue of whether human evolution does show progressive trends, is the subject of this chapter.

The Pattern of Hominid Evolution

The time has come to look in a little more detail at the overall pattern of hominid evolution as shown by the fossil record. As we have seen, the earliest fossils come from a number of sites dated to about 5 million years ago, primarily from sites in the eastern Rift Valley. Until about 3 million years ago these are relatively uniform, and until very recently have all been assigned to a single species – *Australopithecus afarensis*, a name coined by Johanson and White to describe the specimens from Laetoli in Tanzania and Hadar in Ethiopia.[1] These are fossils that show a generally upright stance, but are relatively primitive in other respects: their jaws are protruding, as are their faces as a whole (prognathous), their teeth large, with spatulate incisors and relatively large canines that still protrude above the line of the other teeth. The molars are more distinctly human, but are large in relation to modern forms. The forehead is low and receding and the cranium small with pronounced muscle markings particularly in the nuchal region.

This is a pattern that, with minor variations characterizes a whole genus – the australopithecines (*Australopithecus*) – and *Australopithecus afarensis* represents an early manifestation. When compared with modern humans it is the most 'ape-like' of all the hominids, and is probably as good a missing link as any that is likely to be found. From the neck up it is very ape-like, from the neck down, much more like a human.

However, the suite of material that has been called *Australopithecus afarensis* is itself variable. By far the largest sample comes from the locality of Hadar in south-east Ethiopia, dated to about 2.9 million years ago. The many specimens that have been discovered are all clearly part of this early hominid grade, but there are some differences as well. The differences are partly due to size. Henry McHenry[2] has calculated that the larger of them are 80 kilograms in weight, while the smallest, which include the famous 'Lucy' herself may have been as small as 30 kilograms (and perhaps male and not very Lucylike, as well).[3] It has been argued by some that this range of variation, which exceeds that of the gorilla, is too great to be sexual dimorphism (differences due to sex, usually with males larger than females), and therefore that more than one species is represented. This essentially means that there was a large and a small hominid living in Ethiopia at around 3 million years ago.

This view has also been argued by Brigette Senud.[4] Rather than size differences alone, she has suggested that there are substantial differences in the structure of the arm and the leg of these specimens, with some more clearly bipedal than others. This might suggest that the two species had different ways of life, with possibly one being more bipedal than the other. There may be some support for this view if the foot of Lucy is compared with the Laetoli footprints. The footprints appear to be very much like the stride pattern of a modern human, but the foot of Lucy still possesses a relatively flat sole and a slightly divergent big toe, as is found in other primates. This would all suggest that even by 3 million years ago there was already a divergence of hominid types. However, Johanson and White have strenuously defended their designation of a single species, arguing that fossils, drawn as they are from much larger periods of time than comparative modern samples, usually do exhibit more variation, and

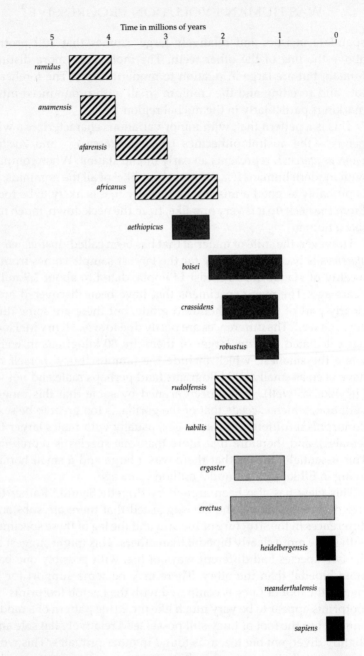

The family of hominids. This diagram shows the main types of hominid (most of which may be thought of as species) that are known from the fossil record, along with when they occur. For most of hominid evolution there have been multiple taxa in existence, and the current situation where there is only one species is of recent origin. These fossil and extinct hominids may be thought of as the species that fill in the evolutionary space between living humans and our closest living ape relatives.

therefore that *Australopithecus afarensis* is a good, single species and the best contender for the basal position on the hominid lineage.[5]

The situation for the early part of the hominid fossil record has been made slightly more complicated recently with the announcement of new species of African hominids.[6] These are *Australopithecus ramidus* and *A. anamensis*. (White and his co-workers have in fact proposed a new genus, *Ardipithecus*, but for simplicity this is not used here.) Both date to between 4 and 4.5 million years ago, earlier than classic *A. afarensis*, but overlapping with the fragmentary pieces that are generally assumed to be part of this taxon. *A. ramidus* is small and ape-like in its dentition and cranial anatomy, suggesting close links with the chimpanzee. Furthermore, unlike other hominids, it has only a thin capping of enamel on its teeth, a characteristic shared with the African apes. This has led to the suggestion that *A. ramidus* is the 'root species' (the word 'ramidus' means root in the local language), and that *A. afarensis* and *A. anamensis* are more derived. Indeed the differences between *A. ramidus, A. anamensis* and *A. afarensis* are excellent evidence that the divergence of hominid forms had already started by over 3 million years ago.

Certainly from about 3 million years ago the fossil record for the hominids becomes more diverse, both geographically and morphologically. Prior to this date all the fossils come from the eastern part of Africa, but after this date they are found in both eastern and southern Africa. The South African sites, which are located in the Transvaal, are those that were discovered by Raymond Dart and Robert Broom in the 1920s and 1930s, but they have been worked systematically since then, and have now yielded hundreds of fossils. Broadly speaking they fall into two groups. There are those that are fairly similar to *Australopithecus afarensis* in having relatively prognathous faces and small and lightly built crania. Overall they are gracile, having thin skull bones and slender limb bones. In size they would have been about 50 kilograms in weight. These were the first early African hominids to be found, and were given the name *Australopithecus africanus* by Dart. Although there have been suggestions that this species occurred in East Africa as well, by and large it is accepted that it is a specifically southern African form.[7]

The other type of hominid found in South Africa is markedly different. Rather later in date, from between about 2 and 1 million years ago, this type is characterized by a much more heavily muscled build, particularly around the jaw and skull. The mandibles are massive, with vastly enlarged molars and premolars. In contrast the incisors and canines are small and cramped. The cranium is low, with a sloping forehead and very rounded head. Most distinctive is the ridge that runs from front to back along the top of the skull. This is the sagittal crest, which indicates the massive muscles that were necessary to move the large masticatory system. These creatures, known as the robust australopithecines (*Australopithecus robustus*) (or by some as *Paranthropus robustus*), have large teeth and small brains, and the whole of their anatomy indicates specializations relating to heavy chewing, most probably of coarse, fibrous plant foods.[8]

This type of hominid is also found in East Africa, from the famous sites of Olduvai and Koobi Fora, and if anything, here they are even more megadontic. Generally they are recognized as another species, *Australopithecus boisei*, but also date from about 2 million years ago. Recently another specimen was discovered by Richard Leakey and Alan Walker, from the west of Lake Turkana in northern Kenya (the Black Skull or WT17000), which was not only earlier (about 2.7 million years old), but also possessed the largest sagittal crest and the smallest brain size of any of the robust australopithecines.[9] Some authorities feel this is just an early version of the robust australopithecines, while others would recognize it as a distinct species – *Australopithecus aethiopicus*. Overall, the robust australopithecines or paranthropines are a diverse group as even in South Africa there is the suggestion that the material, which comes from two sites – Swartkrans and Kromdraai – actually represents two species – *Australopithecus robustus* and *Australopithecus crassidens*.

The robust australopithecines are in marked contrast both with the earlier and more gracile australopithecines, and with another type of hominid that makes its appearance at about the same time. These are the larger-brained hominids. They lack the dental specializations of the paranthropines, having much smaller and more parabolic jaws with more even sized teeth, and have longer and larger brain-cases. Where the australopithecines all have cranial capacities of about 400-500 cubic centimetres, those of the new type exceed 600 cc and often exceed 700 cc. Their faces are small and less prognathous than the other types of hominid, and certainly lack much of the cranial super-structure associated with the heavy musculature found in the robust australopithecines. These are the first representatives of the genus *Homo*, and they are usually placed into a single species, *Homo habilis*. The first examples of this species were found by Louis and Mary Leakey at Olduvai, but other examples are now known from the area around Lake Turkana in northern Kenya.[10]

When *Homo habilis* was first found it was argued that there was insufficient room in the human evolutionary tree for another species, and it was highly contested whether it really was a separate species. Ironically, now that its separate identity has been accepted, it is currently disputed whether what is called *Homo habilis* can really all be accommodated within a single species. Bernard Wood, for example, has argued that if you take two examples of *Homo habilis* such as KNM-ER 1813 and KNM-ER 1470, then you find too much variation. Furthermore, the pattern of variation is not that which is normally found between males and females, so that it is unlikely to be a sexually dimorphic species. He would claim that really what is here is not one species, but two. As well as *Homo habilis* there is another species that should probably be called *Homo rudolfensis*.[11] In addition, by soon after 2 million years ago other types of *Homo* – ones

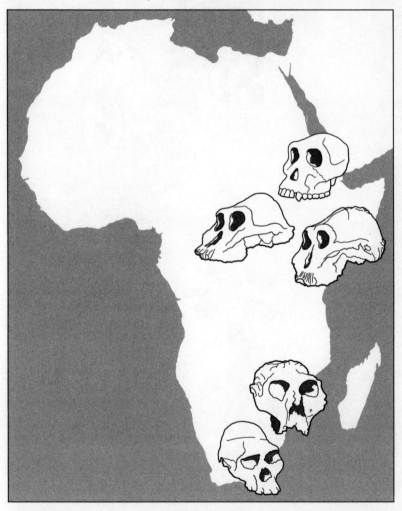

The main species *Australopithecus*. From the top these are *A. afarensis*, *A. aethiopicus* and *A. boisei* in eastern Africa, and *A. robustus* and *A. africanus* in southern Africa.

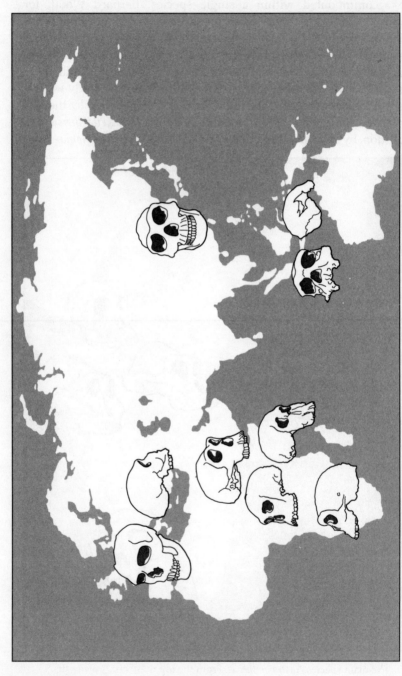

Early *Homo*. The early African species include *H. habilis*, *H. rudolfensis* and *H. ergaster*. *H. erectus* and its descendants are found in eastern and south-eastern Asia, while archaic forms of *homo* occur in Europe (neanderthals, *H. heidelbergensis*) and Africa.

with larger brains and more human bodily proportions – were present, certainly in Africa (*Homo ergaster*) and possibly in south-east Asia as well.[12]

The hominid story is becoming complicated. Taking just the first part of the fossil record, between 4.5 and 1.5 million years ago, the list includes an early primitive form that may or may not be bipedal (*Australopithecus ramidus*), two other relatively ape-like forms (*Australopithecus anamensis* and *Australopithecus afarensis*), and an overlapping but generally later form, *Australopithecus africanus*. These last two may even be further sub-divided into separate species.[13] By 2.5 million years ago, and overlapping in time-range with *Australopithecus africanus*, are the robust australopithecines, and they themselves may constitute up to four separate species. And appearing at about the same time as the robust australopithecines is the genus *Homo*, which itself may consist of three or four species. Most of these taxa are specific to either eastern or southern Africa, and at this time no hominids are known from any other parts of the continent or the world.

The implications of this are striking. Clearly rather than evolving progressively and in a direct line from primitive missing link to modern humans (what is known as anagenesis), hominids branched out and diverged into separate species during their early evolution. The point made in the last chapter that not all hominids were humans becomes rather more obvious when it is remembered that not all these species can be ancestors to living humans, or even be on the line leading to anatomically modern humans. Early hominid evolution is a complex pattern of diversification, and the humans before humanity, to misuse the term somewhat, are clearly an interesting group of animals – so close and yet so far.

Perhaps, though, the story becomes simpler closer to the present. This analysis of the fossil record stopped at about 1.6 million years ago, at which time there were probably five or six extant species of hominid. By one million years ago this number had reduced as most of them became extinct.[14] The known survivor was *Homo erectus*. In contrast to the other hominids, this species was probably larger, and had overall skeletal proportions that were much more like those of modern humans. In chapter 4 it was argued that while all hominids were bipedal in some way, it was only at about 1.6 million years ago, with the arrival of *Homo erectus*, that a type of bipedalism that was essentially the same as that of modern humans occurred. There were still some differences, and *Homo erectus* had much more robust bones and slightly different lower limb morphology than is found in *Homo sapiens*. The differences in cranial structure would have been more striking. Certainly the brain size of *Homo erectus* would have been larger than that of earlier forms of *Homo*, at around 800 cc. As a result the brain case was much longer, and *Homo erectus* had a characteristic long and low cranium. The brow ridges were striking and prominent, and the face still much larger than in modern humans. Overall, while lacking the cresting of the robust australopithecines, *Homo erectus* was still very heavily built and displayed powerful cranial musculature and markings.[15]

This then is the main if not the only representative of the hominids by about a million years ago, the other species having become extinct or having evolved into *Homo erectus*. Early *Homo erectus*, or *Homo ergaster* as some prefer to call it, was, like the other hominids, found solely in eastern and southern Africa, but after at least a million years ago similar forms can be found, bearing the characteristic brow ridges, as far afield as China and Java. Recent dates in both Java and Georgia suggest that this dispersal from Africa to Asia may have occurred closer to 2 rather than 1 million years ago.[16] Moreover, the same basic form persisted until less than half a million years ago, and probably existed in Asia, Africa and Europe (although no clear-cut specimens have yet been found in Europe). They are succeeded by forms that share many of the same basic features, but lack the distinctive long, low head shape and unbroken brow ridges. These are what are usually called the 'archaic *Homo sapiens*' – that is, members of the same species as us (*Homo sapiens*), but still slightly different. They are grouped with *Homo sapiens* on account of their relatively large brains – from 1000 to 1300 cc, close to the figure for modern humans – but still differ in being robustly built and possessing relatively large faces, prominent brow ridges and other features still reminiscent of *Homo erectus*. Here is perhaps a classic example of anagenesis, a gradual and continuous evolution of a single lineage from *Homo erectus* through the archaic forms of *Homo sapiens* to modern *Homo sapiens sapiens*. Modern humans themselves make their appearance at around about 100,000 years ago, but their gradual evolution can be traced in some parts of the world from the archaic types.[17]

It would seem, therefore, that while there are a mass of species in the early part of the hominid fossil record, the later part is much more simple and progressive. It is as if the early hominids were part of a phase of evolutionary trial and error, but that once the winning formula had been hit upon – large brains are better than large teeth – then the path to humanity is straightforward. However, it is probably not that simple. Although continuity can be found, there are also more discrete patterns. Peter Andrews[18] has pointed out

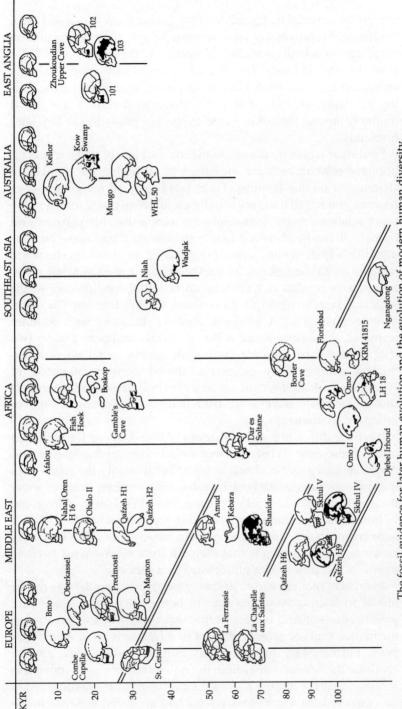

The fossil evidence for later human evolution and the evolution of modern human diversity.

KYR

EUROPE

Bmo
Combe Capelle
Oberkassel
Predmosti
Cro Magnon
St. Cesaire
La Ferrassie
La Chapelle aux Saintes
Qafzeh H6
Qafzeh H9

MIDDLE EAST

Nahal Oren H16
Ohalo II
Qafzeh H1
Qafzeh H2
Amud
Kebara
Shanidar
Skhul V
Skhul IV

AFRICA

Afalou
Fish Hoek
Gamble's Cave
Boskop
Dar es Soltane
Border Cave
Florisbad
Omo I
KRM 41815
LH 18
Omo II
Djebel Irhoud

SOUTHEAST ASIA

Niah
Wadjak
Ngangdong

AUSTRALIA

Keilor
Kow Swamp
Mungo
WHL 50

EAST ANGLIA

Zhoukoudian Upper Cave
101
102
103

that all the material that is called *Homo erectus,* from Africa and Asia, stretching from over 1.5 million years to less than half a million years ago, is actually variable. In particular the Asian forms differ from the African forms. This in itself is not so surprising, given the space and time involved. What is surprising though is that, according to Andrews and others, the characteristics that are found uniquely among the Asian *Homo erectus* are not found in any later hominids.

Evolution works by the accumulation and loss of traits, and evolutionary relationships are measured by assessing the number of shared and unique features. Traits can be shared for a number of reasons, and not all traits are equally useful when trying to determine exact relationships. For example, in comparing chimpanzees and humans it can be observed that both possess a backbone, but this does not help determine how closely related these two species are in relation to, say, monkeys, as backbones are common to all vertebrates. This is what is known as an ancestral condition, or in the jargon, a symplesiomorph. In contrast, neither humans nor chimpanzees possess a tail, whereas monkeys do, along with all other primates. Possessing a tail is the ancestral condition, and so two species that share the characteristic of having no tail show a distinctive pattern. They possess a shared derived character, or synapomorphy. In this case, because the condition of 'tail absence' is something new it is a characteristic that says something about evolutionary relationships.

It is these different patterns that can be found among the hominids of the Pleistocene. What has been called *Homo erectus* in Asia and Africa certainly shares characteristics, but it is only the Asian ones that possess unique derived features. Furthermore, none of those derived features, which relate to the details of skull morphology, can be found in later hominids, those belonging to what have been called here archaic *Homo sapiens* and modern *Homo sapiens*. The inevitable conclusion to be drawn is that the path from *Homo erectus* to *Homo sapiens* is not quite as straightforward as it seemed.

It appears that the early African *Homo erectus* populations can be linked to later archaic hominids in both Africa and Europe, and possibly parts of Asia as well, by these apomorphic traits. In contrast, the unique traits or apomorphies of true *Homo erectus* in Asia disappear. Furthermore, fully modern humans, when they appear, continue the African rather than the Asian characteristics. What this means is that the middle part of the evolution of *Homo* consisted of two lines at least. Once *Homo erectus* had appeared in Africa it dispersed out into Asia and possibly Europe as well. The Asian line,

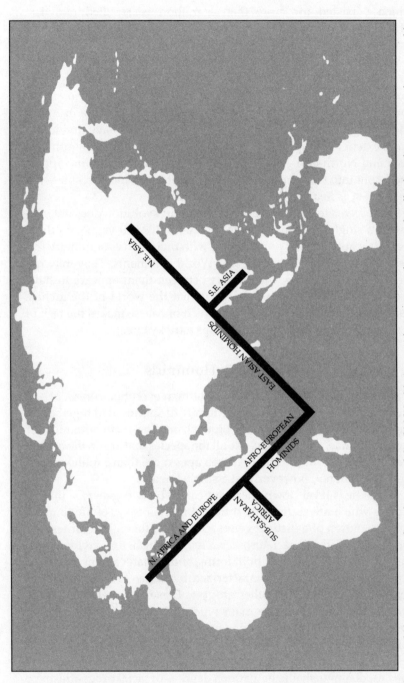

A phylogeny of *Homo* superimposed on the map of the Old World. As populations spread geographically they diverge in evolution, and hence it is to be expected that there should be some geographical conformity to the evolutionary relationships of hominids, as is the case.

which persisted for more than a million years, died out. The Afro-European line, though, diverged into a number of regional populations, one of which became modern humans.

What this suggests is that far from being a simple linear process, later hominid evolution consists of the same pattern (although less extreme) of divergence as the early part. Ancestral *Homo erectus* spread from Africa and formed two separate groups, one in Africa and Europe, one in Asia. The Asian line may itself have divided to some extent geographically, with distinct populations in South-east Asia and North-east Asia. Similarly the Afro-European line apparently split into two populations, one in Europe, which became what are known as neanderthals, and one in Africa.

What seems to be the case is that hominid evolution does not consist of a single evolving lineage. By about 200,000 years ago there were probably distinct populations, with their own evolutionary trajectories on each of the major Old World continents. They may not have been full species, but contact between them appears to have been limited. The simple message is that the world of the archaic hominids was a complicated one, which of course makes the task of sorting out evolutionary relationships a tricky one.

Too Many Hominids

What, then, does the fossil record of human evolution consist of? At the broadest level there is a single family of species, tied together by a number of traits, particularly bipedalism. These are hominids, or the Hominidae, and they include all the species that are on the branch that diverged from the other African apes more than 5 million years ago. They are not, however, just a single lineage.

At the next level down there appear to be two genera – that is, groups which are sufficiently different in their ways of life to merit the recognition of a distinct genus for each of them. The earliest are the australopithecines (*Australopithecus*), which are divided into early and relatively lightly built forms, and the later and more robust type of *Australopithecus*, characterized by large chewing teeth and heavy musculature. The other group is *Homo*, with relatively large brains. These are the three main types of hominid, or put another way, the three things that it appears possible to do with bipedalism.

However, even within these groups there is variation, as we have seen. *Australopithecus* may consist of as many as four different species, distinguishable by various degrees of facial prognathism or body size. The robust australopithecines may also consist of four species, two in East Africa, two in South Africa, all except one of them

(*Australopithecus aethiopicus*) broadly contemporaneous. *Homo* is a rather more complicated group to sort out. There are probably two early forms, with relatively small brains, but varying in the size and shape of their faces. Then there is what has been called *Homo erectus*, which also should perhaps be divided into two separate groups, an Asian one and an African one. After this are all the forms that have been referred to as 'archaic *Homo sapiens*'. It has generally been accepted that they all belong to the same polytypic species as ourselves, but Ian Tattersall[19] and Clark Howell[20] have recently suggested that they should really be accepted as separate species on the grounds that they have unique characteristics, are radically different from modern humans, and appear to be evolving independently. The exact number is difficult to determine. Many might certainly recognize *Homo neanderthalensis* as a distinct group of hominids living in Europe and parts of the Middle East during the last 100,000 years. Their precursors, the first Europeans, from between half a million and 200,000 years ago (*Homo heidelbergensis*), would be separate species as would the African and Asian 'archaics'. These last appear to be the ones that resemble modern humans the most (a problem to be explored in more detail in the next chapter), and it has been argued that they are the only group of the 'archaics' that are actually on the line leading to modern humans. Finally, modern humans, according to this scheme, would no longer be *Homo sapiens sapiens* (i.e. a sub-species), but *Homo sapiens*, a separate species.[21]

As can be seen, even at this simple level the accountancy of hominid diversity is not easy. A minimalist view would be that there are at least eight species, but at the other extreme as many as seventeen would be recognized if all the distinctions discussed here were accepted. Wood has made a good case for thirteen species, which fits well with theoretical expectations.[22] The precise number will never be known, and anyway the species concept is almost certain to break down when temporal variation is taken into account. On the one hand we can argue that the fossil record is incomplete, and therefore there are still more species out there to be discovered. On the other hand, it might be argued that the anatomical minutiae have been overinterpreted, resulting in too many spurious distinctions, and that the number is more likely to be towards the lower figure of eight. However, whichever is the case there remains a problem and a challenge. The problem is that the human evolutionary tree cannot be reconstructed as a simple line – a sort of evolutionary conifer. It is clearly a branching bush, sprouting stems all over the place. This means that actually determining who is most closely related to whom, and who is ancestral to whom, is a major technical problem. Here this will be explored only briefly. The more major challenge is to start thinking of human evolution in a different way. Instead of a ladder with humans as the pinnacle, there is a bush with humans as just one little twig. More startling, the twig of modern humans is in many ways quite a departure from the rest of the hominids, and occurs only relatively recently, as the previous chapter showed. What is the meaning of human uniqueness when there is such a plethora of hominids?

The Evolutionary Ladder and the Fallacy of Progress

The abundance of hominid types has come as a surprise to many. That it was a surprise perhaps reveals hidden assumptions about how evolution works. Since Darwin, the general model has been that of the *scala naturae*, the scale of nature. This pictures animals as a series of progressive developments, the one replacing the other as they scale new heights of evolution. Multicellular organisms are a progression from single-cellular ones, vertebrates are a progression from invertebrates, warm-blooded animals are a progression from cold-blooded ones. Naturally, in this progression, humans – or more objectively, intelligent social beings – are a progression from unintelligent, solitary ones. Humans represent the next or most recent stage of evolutionary progression. It is only natural, given the complexity of the human world, to place them at the top of the evolutionary scale, and that evolution should be seen as a ladder leading to ourselves perched, albeit precariously, on the top rung.

The problem is that while there may be directional trends in evolution that we can observe from the vantage point of hindsight, there is a logical error in inferring from this that the mechanism of change is also a progressive one. Species arise and are then replaced – for example, the dinosaurs by the mammals – and we infer from this that the one represents a progression from the other. The tendency of evolution to wipe out most of its creations through extinction buttresses this idea. However, the actual pattern of evolution at the small scale is very different, and the process rather contrary to the idea of a ladder.

This view emerges by looking at another group of animals. The Bovidae, or bovids, are the two-toed ruminants with horns.[23] They include the antelopes, the waterbucks and reedbucks, the eland, the buffalo and even the domestic cow. Although primarily African in distribution they can be found widely in all the continents except Australia. The first observation is that they are very diverse – there are many genera and many species, ranging from the very small dikdik to the massive African buffalo. They are linked together in evolution by a number of features, such as their horns (as opposed to the antlers of the deer family, the Cervidae). One would be hard pressed, though, to look at them and view their evolution as a progressive ladder from one group to another. Are the waterbucks at the bottom and the eland at the top? Or perhaps it is preferable to say that the thick lumbering buffalo is the most primitive and the graceful gazelles are the most progressive. Clearly, there is no basis for making this decision, and we have only to look at their characteristics to realize how inappropriate it is to approach evolution in this way. The eland and the buffalo are different because they do different things and live in different environments. The buffalo is a large,

A cartoon version of the *scala naturae* – the idea that evolution is a progressive process leading to humans at the end. The correct interpretation, as Darwin himself realized, is divergence and adaptive uniqueness rather than progressive trends.

unselective grazer, whereas the eland feeds in light woodland on a mixture of grass and leaves. The oryx is a specialist in living in very dry, desert conditions and can survive with virtually no water, whereas the waterbuck is highly water-dependent, feeding off lush grass. One of the Reduncinae, the group to which the waterbuck belongs, is the puku, and this has specializations in its feet, which are splayed, allowing it to walk on soggy, marshy ground. The characteristics of the bovids are markers not of an evolutionary progression, but of the process of adaptation to the needs of a wide variety of habitats and food types.

There is nothing very surprising about this, and we can track in the fossil record the appearance and diversification of these various groups. Evolution among the bovids is, like the hominids, not a single line, but a spreading out in bush-like form of all the varieties we see today, and along the way numerous others have disappeared. Because we do not want to rank the bovids in a hierarchical and progressive fashion, we are happy to allow them to diversify into a range of equal types. Not so with apes and humans. Rather than seeing bipedalism as an alternative to quadrupedalism, people prefer to think of it as an advance. Rather than asking why brain size might be a way of adapting to different habitats, it is easier to build a hierarchical system from the smallest, and most primitive, to the largest and most advanced. The fossils of the whole family, the Hominidae, similarly mark out this progression. The surprise of the fossil record is that this is not what happens. *Australopithecus robustus*, with a relatively small brain, was contemporary with the larger-brained *Homo*, if not slightly later. Neanderthals, who are at least contemporary with modern humans, have equally large brains. The direct evidence of the fossil record shows that the progression is one that we impose with hindsight. It is not, though, the way that evolution works.

The surprise, then, reflects prejudice, or perhaps the ignorance palaeoanthropologists had of the evolution of other groups. Most groups show a pattern of diversification, particularly early in their evolutionary history, and there is no reason why this should not be the case with human evolution as it is with the majority of other species. Evolution is not a ladder, to use the words of Gould, but a bush.[24] The task is not to map out a progression, but to look at the shape of that bush, to see where it is thick, where it is straggly, and to draw inferences about the range of adaptations that are represented. Different species are alternative responses to multiple environmental conditions. The real interest in human evolution lies in looking at what those conditions are, and in trying to understand not why ancient hominids did or did not evolve into modern humans, but why they took the form they did. These 'humans before humanity' are the bush, and modern humans just one among many potential and actual end points. Furthermore, in some cases they are not just the humans before humanity but the hominids alongside humanity.

The Hominid Radiations

The story of hominid evolution can be told in a number of ways. The most traditional is that of a number of stages. We have seen that by and large this is inappropriate, as the picture is not just one of successive changes. None the less there does remain more than just an element of truth in the view that evolution is about the creation of novel forms. New things do appear from time to time, and it would be misleading to ignore this. We can still map the appearance of traits, but we must be careful not to then build evolutionary trees on this structure. When it is looked at as a series of traits then we can recognize a number of time-successive events – evolutionary trends.

The first is the adoption of bipedalism, probably some time before 4 million years ago. The first 3 or 4 million years of hominid existence were probably characterized by a number of species of these essentially ape-like bipeds. From some time before 2 million years ago, other specializations began to appear, but these showed *two*

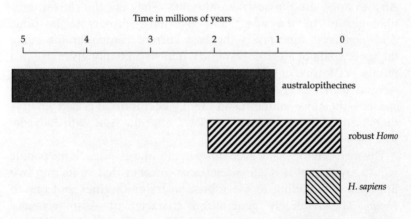

Grades in hominid evolution. Although evolution is largely about divergence, adaptive trends do occur such that certain groups of species can be said to share the same level of organization and adaptation. In hominid evolution it is likely that the australopithecines represent a grade best thought of as bipedal apes, the early *Homo* as more encephalized but with a robust skeletal architecture, and modern humans as showing a trend towards greater skeletal gracility and behavioural flexibility.

rather than a single trend. One was the appearance of dental specialists, those creatures with large cheek teeth and powerful chewing
muscles. The other was that represented by *Homo*, showing brain
rather than dental enlargement. Associated with this may have been
a number of marked behavioural changes, including the incorpora-
tion of more meat in the diet and an increased use and manufacture
of tools. *Homo* was also the first hominid to be more than a part of the
African fauna alone. The final major event was the appearance of
anatomically modern humans, marked by much more lightly built
skeletons, higher and more rounded crania, reduced faces and a
whole suite of new behaviours, often indistinguishable from those
we find today. For those looking for the progressive trends of
hominid evolution, these are they.

Another way of looking at hominid evolution is phylogenetically,
the reconstruction of precise evolutionary relationships. This is probably the most technical of all the perspectives that are possible, a
question of matching the myriads of anatomical features, weighting
different characteristics, and sorting out chronological and strati-
graphical relationships. The fine details of these would make a book
on their own, but it is not the intent of this one to explore them in
detail. An evolutionary tree can be constructed linking the various
fossil groups together, following the implications of the taxonomy
presented earlier.

At the root of the tree, subsequent to the divergence with the
African apes, are the australopithecines. They are the closest morphologically to the apes, and the most proximate in time.
Australopithecus ramidus is the best current contender for being
the sister clade of all later hominids. It most probably gives rise to a
number of forms, of which *Australopithecus anamensis* and *Australopithecus afarensis* are eastern African forms, and *Australopithecus
africanus* the more southern and slightly later form. It is also possible
that *Australopithecus afarensis* persisted after the split with *Australopithecus africanus*.

The next phylogenetic event is open to some dispute. Some people
would argue that it is *Australopithecus africanus* that splits into two
lineages, one leading to the robust australopithecines and one to
Homo. The relatively generalized character of *Australopithecus
africanus* would seem to allow for this. However, some scientists feel
that *Australopithecus africanus* is already too specialized dentally to be
the ancestor of *Homo*, and is later than the split of lineages and solely
an ancestor of the robust australopithecines. Others argue that it is
already too late to be their ancestor in view of the early date (2.6 million years ago) for one of them, *Australopithecus aethiopicus*. Indeed, it

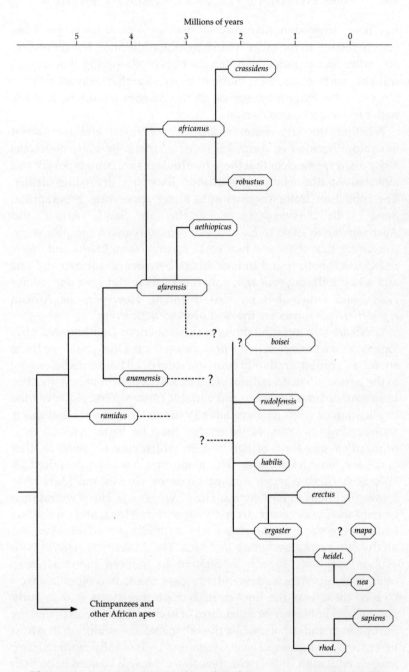

Millions of years

There is no more contentious subject than the reconstruction of hominid evolutionary relationships, and the one presented here is one possible among many that are equally plausible. Few phylogenetic reconstructions survive for long; the above has been selected to emphasize the multiple events and their biogeographical basis in hominid evolution.

has been suggested that *Australopithecus aethiopicus* is not even closely related to the other robust australopithecines, but represents an earlier and parallel evolutionary trend.[25] Resolving this issue is fraught with difficulties. Suffice it to say that *Australopithecus africanus*, the longest known of all the African hominids, remains both pivotal and controversial.

Whether the last common ancestor of *Homo* and the robust australopithecines is *Australopithecus afarensis* or *Australopithecus africanus*, it seems clear that these two lineages are contemporary and alternative paths of hominid evolution. Each itself diversifies further. The robust australopithecines split along apparently geographical lines, with *Australopithecus robustus* in South Africa and *Australopithecus boisei* in East Africa.[26] *Homo* is also a complex story. Accepting that there are two early forms, *Homo habilis* and *Homo rudolfensis*,[27] both found in East Africa between about two and one and a half million years ago, only one of them can give rise to later *Homo*, and which it is remains disputed. However, the African *ergaster/erectus* forms are the next phylogenetic event.

There are two interpretations of subsequent or later hominid phylogeny, as we have seen in the previous section. One would see *Homo erectus* as a broad, gradually evolving stem of all hominids belonging to the period 1.5 to 0.5 million years ago. This in turn would give rise to an equally broad-ranging and variable *Homo sapiens*. However, the implication of what we have already discussed suggests that this is an oversimplification. *Homo erectus* may be better viewed as a primarily Asian form of hominid, in which case we need another name for the African form. The name that has been proposed by Wood[28] is *Homo ergaster*, a name given by Groves and Mazak[29] to some specimens in East Africa. The phylogeny of *Homo* would then be read as *Homo ergaster* diversifying within Africa, and on the one hand giving rise to *Homo erectus*, which rapidly spread into Asia and on the other hand persisting in Africa. The Asian *Homo erectus* is an evolutionary side branch in relation to modern humans. *Homo ergaster* and its African descendants gave rise to two other lineages. One of these was the lineage in Europe that starts with a fairly generalized but larger-brained form of hominid, and evolves into the European neanderthals, while the other sees the evolution in Africa first of archaic forms of *Homo sapiens* and then of fully modern *Homo sapiens*. Although in later chapters this view will be seen to be both controversial and more complicated, it can be argued that this African species then dispersed throughout the world, replacing, with more or less interbreeding, the other and more archaic forms of hominids.[30]

Such phylogenetic reconstructions can be more or less specific. They are also best viewed diagramatically, for this makes clear the branching sequence and nature of evolutionary events. It is phylogenetic reconstruction alone that shows the path from a direct ancestor through to modern humans. Such a path does exist, but the message of this chapter is that this is just one among many evolutionary paths within the Hominidae.

There is one further way of viewing hominid evolution, and one that is germane to the idea of there being many chronologically overlapping hominids. The example of the bovids shows that rather than rising progressively in a single evolutionary direction, they radiated out into different ecological niches. This is the typical way in which lineages evolve. The term used in evolutionary biology for such an event is that of an adaptive radiation. The concept underlying this is that species that are closely related to each other – that is, that have a common ancestor – none the less show differences. These differences, though, are all in the form of variations along a theme. An adaptive radiation in evolution is the diversification of groups of plants or animals that show these variations based on a common ancestral trait. For example, the bovids are an adaptive radiation. They are linked not only by a common ancestor but also by a series of common adaptive characteristics – in their case, the ability, through their ruminant stomach, to process a large number of plant foods. The variation comes when we look at the different contexts in which bovids have adapted themselves to use this basic ability. There are the dikdiks and duikers, as very small, forest-dwelling specialists; there is the puku, with its ability to pursue this strategy in the marshes of Africa. There is the oryx in the desert, the klipspringer among the rocks of Tanzania, and the buffalo as the most generalized mowing machine. Each of these is an adaptive variant on the basic theme of the tropical ruminants.

Cats are an even clearer example. All cats have a common ancestor, and what they have all inherited from this common ancestor is the ability to stalk and ambush prey. This contrasts with the dogs, for example, who run down and harry their prey. The stalk/ambush strategy, with its suite of anatomical and behavioural adaptations, links all cats, and it has been the basis for a massive radiation over four continents; in body size, from the lion to the Scottish wildcat; in habitat, from the forests of Amazonia to the sub-Arctic; and in prey, from wildebeest to small rodents. Each species is an adaptive variant of the basic strategy, evolving to fit specific conditions and take up new opportunities.

This concept can be used to understand hominid evolution. All hominids share a unique ancestor, and have diverged from this. The fossil record seems to suggest that what that common ancestor had, and which is the common characteristic of all hominids, is bipedalism. The hominids, it appears, are the apes that walked upright. However, rather than just leading in one direction, bipedalism seems to have opened up opportunities in a number, and consequently there was an adaptive radiation of these bipedal apes to pursue the strategy of bipedalism in different ecological contexts. One such path was indeed the genus *Homo*, with its large-brained variant; the other was the robust australopithecines with their megadontic specializations, while the other australopithecines appear to be more generalized earlier forms.

Hominid evolution, therefore, can perhaps also if not better be seen not as a ladder, nor even as a bush of phylogenetic branches, but as a series of adaptive radiations taking place over the last 5 million years. The first of these were the basic bipedal apes, confined to the

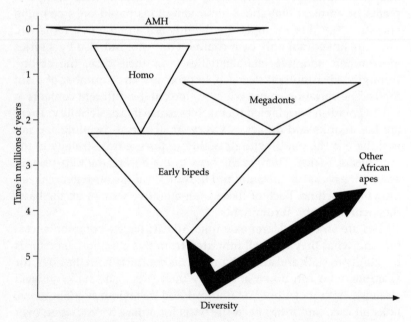

Hominid evolution seen as a series of adaptive radiations. The first is the radiation of the early bipedal apes (australopithecines). The second is the specialization of some of these on coarse vegetable foods, resulting in megadontic adaptations. The third is *Homo* becoming more encephalized, diversifying and spreading out of Africa. The last is the dispersal of modern humans. This last has geographical separation with no major evolutionary divergence.

drier parts of Africa, and in turn giving rise to two further adaptive radiations – *Homo* and robust australopithecines. Each of these adaptive radiations has its own specific adaptive theme – brains and teeth respectively – and its own range of diversification. The morphological range among the paranthropines is great, but the adaptive variation among them is probably relatively small. In contrast, *Homo* radiated rather less morphologically, but enormously ecologically as it spread from Africa, giving rise to distinct geographical populations.

This flowering of hominid types is the raw material of evolution. Had we been there at the time we would have been hard pressed in all probability to pick out which ones were going to succeed and which were doomed to extinction – for each one was adapting to local conditions, some of which made other things possible, others of which acted as major constraints on subsequent evolution. This is the true process of evolution – an endless production of novel ways of doing things, exploring alternatives, trying out new strategies as conditions themselves shift and change all driven by natural selection. The blinkers of hindsight make us want to read this as a pattern of progress, a single arrow passing through time. In contrast, though, by looking at the events of hominid evolution as a series of adaptive radiations it is possible not only to trace the story of direct ancestors, but also to begin to make sense of both their contemporaries and the conditions that produced all of them.

Uniqueness Reconsidered

This chapter started with a notion of evolution as a process of gradual change through time, with lineages evolving in particular directions. It has ended instead with a model of bushes. These bushes can be looked at in fine focus to determine precise evolutionary relationships, or else can be examined from the more distant perspective of the shape of the bush rather than the details of the tree. Each approach to evolution is a valid one, depending on the type of question asked. The question of progress in evolution has been considered and it was suggested that while it is possible to trace the progressive appearance of novel features, this does not, in the context of what else is happening, constitute simple progressive change. Evolution is not centrally about progression, but about diversification. In this context hominids appear to be like any ordinary mammal in their evolution, starting as a slim stem and then radiating into a number of distinct branches. These are the adaptive radiations of hominid evolution.

We can, though, ask one further question about this pattern – where does it leave human uniqueness? Earlier chapters showed that it was the uniqueness of humans in relation to the living apes and monkeys that acted as a barrier to truly comparing humans and the rest of the biological world. There appeared, in other words, to be too great an evolutionary distance between ourselves and the apes. However, perhaps more significantly than anything else, the fossil record shows that this evolutionary space was not empty in the past. It was in fact filled with a number of other hominids – possibly as many as seventeen of them, although not all at the same time. That they are not there now is a product of other ecological and evolutionary factors relating to extinction – it is not an inherent and unbridgeable gap. Had the robust australopithecines survived in some pocket of East Africa, or had the neanderthals held on in Siberia, or *Homo erectus* in Java, we would not be struck by the difference between humans and chimps, because we would be able to see the other hominid species that fill in the gaps. Moreover, it is now known that they filled in the gaps not merely as temporary stepping stones on the way to being human, but as viable species in their own right, some of which lasted as long as a million years. If nothing else, the fossil record, by virtue of its amazing diversity, shows that the uniqueness of humans is a product not of the evolutionary process of speciation, but of that other one, extinction.

The problem of extinctions is one of the most fascinating and poorly understood in human evolutionary biology. However, another question has been pushing itself forward during the course of this chapter. At all points Africa has lain at the core of these discussions. The persistence of Africa as a centre of evolutionary novelty is clearly an interesting issue.

6

Why Africa?

The Human Race

Evolution is always associated with time. After all, evolution is a process that occurs through time, and it is the extraordinarily long stretches of time that are involved that strike the imagination. Dinosaurs lasting 100 million years or hominids evolving over 5 million years, these are the things that make evolution different from other branches of science or from everyday life. We could easily ask the question of whether there are some periods of geological time that are more interesting than others. Presumably such a period would have a large number of evolutionary events taking place – new species arising or mass extinctions, for example, and to all but the most dedicated palaeontologist these periods would no doubt be more compelling than an aeon where everything stayed the same (this is of course rather unfair, as from a strictly scientific point, explaining why there is no change should be equally as important as explaining why there is change, but human enquiry is not necessarily fair). We have already seen in chapter 3 that an examination of hominid evolution through time yields up some interesting patterns and some surprising insights into the question of when exactly we became human.

However, time is not the only dimension in which evolution occurs. All organisms have to be somewhere, and their geographical location is an important element of their evolutionary context. By and large evolutionary events do not occur over very large areas, although the consequences of such events may spread far and wide. Rather, evolution occurs in small or specific geographical pockets. This observation permits a number of questions that are of general interest as well as of more particular importance to problems in human evolution. In the same way that evolutionary events are not evenly distributed through time, so too are they not distributed evenly across the planet. There is no grand Darwinian quota system that ensures that Britain or America get their fair share of new species or are not picked out for an undue proportion of extinctions. There appear instead to be evolutionary hotspots, places where evolutionary events tend to occur relatively frequently. In looking at hominid evolution we have seen that the continent of Africa is recurringly the place where things seem to happen. This simple observation can be used as a springboard for looking at the geographical patterns of human evolution, and asking not only why Africa is so important, but why there is a geographical pattern to human evolution at all.

This is a question that others have addressed from a rather different standpoint. Louis Leakey,[1] perhaps the greatest champion of the 'African perspective', was particularly concerned with this issue. Being an African himself he was convinced that Africa was the cradle of humanity, contrary to most of his contemporaries. Some, such as Weidenreich[2] and Koeningswald,[3] thought Asia a more likely homeland for humans, either on climatic grounds – no advances could be made in very hot climates; or on historical grounds – Asia was the continent with the greatest and oldest civilizations. Others preferred Europe, largely on the basis of chauvinism. The French were happy to see the Neanderthals and the glories of the cave art of the Dordogne as the natural proof that the French had always led the way; while the English, proud possessors of the (unfortunately fraudulent) Piltdown man, felt it quite acceptable that the origins of humanity should be not just in England, but in the Home Counties, the natural habitat of all good and progressive things.

Leakey took up the gauntlet on behalf of Africa. In an intellectual environment that saw Europe as the advanced continent and Africa as the laggard, he was determined to show that this had not always been the case. Africa, he wanted to show, had previously been the leader, even if it was now taking a well earned rest after 2 million years of pacemaking. This view of evolution as inextricably linked to historical development and consisting of a race in a single direction was deeply held, as was the view that Africa was the loser. What with hindsight is interesting about Leakey's stand is that he obviously accepted this view, and differed only in his interpretation of the order of the runners and the length of the race. He did not reject the whole notion that some parts of the world were leaders and others laggards.

What has been learnt so far in the preceding chapters, though, is that evolution is not unidirectional, and that it is not a global race. Evolutionary events occur because of the conditions under which organisms find themselves locally. They are not 'going anywhere', but are adapting, and among hominids this appears to be a diverse and complicated process. The question 'why Africa?', therefore, is not about why Africa was first (or indeed last, if that is the case), but why different events seem to occur at specific places as well as at specific times. We are certainly interested in asking the question 'where did we become human beings?' but not for the sake of handing out medals. Because on the large scale hominid evolution is not progressive, to look for a progression is an unsuitable approach. Instead the key question is what sort of evolutionary conditions promote or inhibit or simply affect the workings of evolutionary mechanisms.

Just an African Ape?

Humans are African apes. Jared Diamond,[4] in a recent book, has gone so far as to describe humans as 'the third chimpanzee' and to place humans in the same genus as the common chimpanzee and the pygmy chimpanzee. This is probably unwarranted, as genera normally denote a common adaptive basis, and we have seen that hominids with their distinctive bipedal gait are significantly different from chimpanzees. None the less, the point is a good one. As we saw in chapter 3, humans are linked specifically to the African apes, not to the great apes as a whole. Genetically the closest similarities among the great apes are to be found among humans and chimpanzees. These two groups may therefore be thought of as, in the jargon, sister clades or branches. They have a recent and unique common ancestor. That common ancestor in turn has as a sister clade the gorilla, the other African great ape. It is this pattern of relationship that serves as a basis for saying that humans are African apes, for they represent a distinct group, relative to the other great ape, the Asian orang utan. Studies in molecular biology would seem to show that geography, not superficial morphology, is the key factor in the evolutionary relationships of the apes, and as such we belong to the African part of it.

As discussed earlier, this was a view held by Charles Darwin[5] himself, although most subsequent authorities saw it differently. Darwin used living animals that he thought represented the most

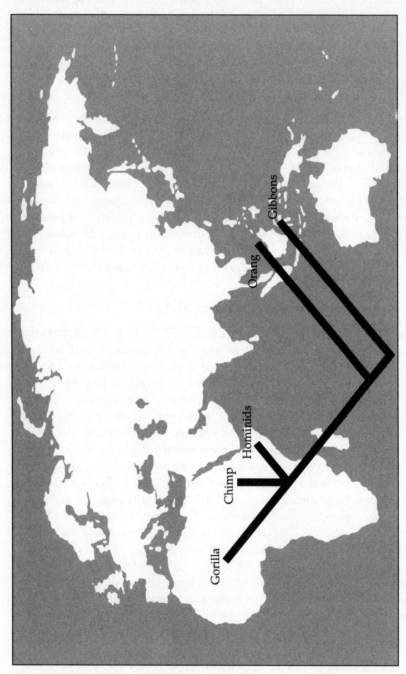

Evolution recapitulates geography: evolutionary relationships of the apes shown in relation to a map of the world.

probable 'ancestor' of the human lineage, whereas other authors placed more emphasis on the geographical patterns of subsequent rather than antecedent events. The fossil record that was examined in detail in the previous chapter would seem to bear this out. All the known species of hominid prior to two million years ago are uniquely African (and specifically sub-Saharan). The australopithecines do not occur outside Africa, and neither do three of the species of *Homo*. Furthermore, if we accept the multiple species of hominid described in chapter 5, then of the seventeen types of hominid known, twelve had their origins in Africa, three in Asia and two in Europe. Looked at another way, there were probably at least two and probably three major geographical radiations in hominid evolution. The first of these occurred between two and one million years ago and was from Africa to Europe and Asia, and the second was probably less than 100,000 years ago and was again from Africa to Eurasia, Australia and the New World. There may well have been other dispersals, such as the European neanderthals into the Middle East and North Africa, or from Europe into Asia by hominids about half a million years ago, but the basic pattern is 'out of Africa'.[6]

There is clearly justification in saying that hominid evolution is an African affair, with only small walk-on parts for the other continents. Africa is the primary donor of novelty, Europe, Asia and eventually Australia and the Americas the recipients. While this may cause twinges of chauvinistic pride or despair, the problem from the point of view of evolutionary biology is to explain this state of affairs. Is there something special about Africa or is it just a coincidence? If there are special factors operating in Africa, are they related to the climate or the environment, or is it a matter of the types of primates that happened to be there? Is it possible in evolution to talk about centres and peripheries, and if so, what are the evolutionary implications of being in the middle or on the edge? Or more drastically still, perhaps it is all just a geological accident. Africa is not so much a good place for hominids to evolve, but a good place for them to die and hence be fossilized. There is an element of all these factors in the Africanness of the human species.

Geological Accidents and Evolutionary Hotspots

When we look at the fossil evidence for hominid evolution we do not find a nice even spread across the planet. Continents such as the Americas, Australia and Antarctica yield either no or only very recent (in evolutionary terms) specimens. This is to be expected as they are far from the areas where humans evolved, and have only been colonized in the later phases of prehistory. Although the dates vary considerably, the areas of the 'Old World' – Africa, Europe and Asia – all have a record of human fossils stretching back at least a million years, and of course, in the case of Africa, over 5 million years. This continental distribution, though, can be misleading. When we look at the fossil evidence in detail, we do not find hominids everywhere across Africa. They occur instead in relatively small pockets in eastern and southern Africa. More precisely, the sites are located in two distinct geological features.

The first and most dramatic of these is the Great African Rift Valley. This is one of the geological wonders of the world. Essentially it is a line of tectonic stress – that is, where two plates of continental matter are tearing apart – that runs from the Luangwa valley in Zambia up through Tanzania, Kenya and Ethiopia to the Red Sea, where it divides and continues on the one hand out into the Indian Ocean and on the other along the Red Sea and into the Jordan Valley. Over the last 20 million years or so the areas to either side of this fault line have been moving apart, and as they have done so, the land in between has dropped down into a valley. In fact it is not a single valley, but a mass of smaller fault scarps where the land has moved up and down. It has created a complex landscape of cliffs and lake basins, dotted with volcanoes. It is among the most active terrestrial landscapes in the world. It is within these lake basins that the best hominid fossil record lies. The fossils are found in the lake and river sediments that have built up. Because of the extensive tectonic movements these have been piled high, and then subsequently exposed by further movements of the earth. The gentle build-up of the lakes, within which animals are buried and preserved, is ideal for fossil formation, while the constant shifting of the strata makes it easy for the palaeontologists subsequently to find them eroding out of the scarps.[7] Amongst these are the hominids of the last 5 million years.

The other fossil-rich area is in the Transvaal of South Africa.[8] This is a very different geological context, consisting of ancient limestones that stretch back to the Mesozoic. These limestones have in places been dissolved out into subterranean caverns, some of which have been exposed by erosion. Into these caverns and holes have fallen the detritus of the millennia – rocks and mud and the bones of animals, including hominids.

The point here is that these are areas that are not necessarily good places to live and to evolve, but good places to die. (Good, that is, if you want to become a fossil.) Far more is known about these regions in the past, not necessarily because these areas were richer in ancient hominids and other animals, but because they preserved them in

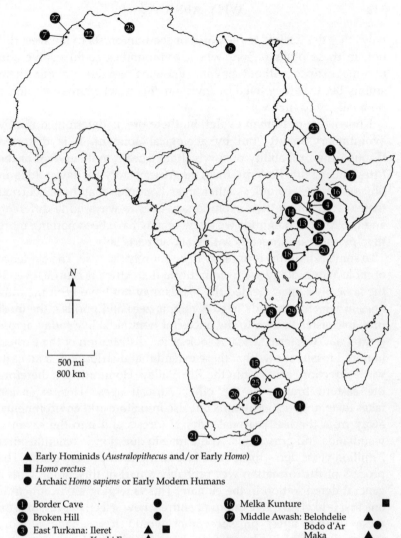

▲ Early Hominids (*Australopithecus* and/or Early *Homo*)
■ *Homo erectus*
● Archaic *Homo sapiens* or Early Modern Humans

① Border Cave	●	
② Broken Hill	●	
③ East Turkana: Ileret	▲ ■	
Koobi Fora	▲	
④ Fejej	▲	
⑤ Hadar	▲	
⑥ Haua Fteah	●	
⑦ Jebel Irhoud	●	
⑧ Kanapoi	▲	
⑨ Klasies River	●	
⑩ Kromdraai	▲	
⑪ Laetoli	▲	
⑫ Lainyamok		
⑬ Lake Baringo: Chemeron	▲	
Chesowanja	▲	
Tabarin	▲	
⑭ Lothagam	▲	
⑮ Makapansgat	▲	

⑯ Melka Kunture	■	
⑰ Middle Awash: Belohdelie	▲	
Bodo d'Ar	●	
Maka	▲	
⑱ Olduvia Gorge	▲ ● ■	
⑲ Omo	▲ ●	
⑳ Peninj	▲	
㉑ Saldanha	●	
㉒ Salé	■	
㉓ Singa	●	
㉔ Sterkfontein	▲	
㉕ Swartkrans	▲ ■	
㉖ Taung	▲	
㉗ Thomas Quarries	■	
㉘ Tighennif	■	
㉙ Uraha	▲	
㉚ West Turkana: Lomekwi	▲	
Nariokotome	■	

Map of Africa, showing the main fossil sites.

ways that the deserts to the north or the rainforests to the west did not. In these other regions whole communities could evolve and become extinct without anyone knowing anything about them simply because they lived in environments in which fossils stand a poor chance of survival.

Knowledge of human evolution, therefore, is determined not by evolutionary reality, but by geological accident. It is distorted through the probability of survival after death, not survival of the fittest during life. Perhaps, then, the African story we have told is an illusion. Across countless other areas hominids may have scurried their way through their evolutionary lives in ways quite different, and the African hominids we know of may have been nothing more than peripheral characters with walk-on parts only.

To some extent this is unlikely to be entirely the case. The evidence of molecular biology seems to indicate that Africa is central, even if the areas that have yielded the fossils may not have been the only ones in which hominids lived. Chimpanzees and gorillas, the most probable sister branch to the ancestral hominids, live today across central and western Africa. A look at the distribution of the earliest hominid fossils shows that these extend that distribution in an eastward direction, across into the Rift Valley. Hominids are therefore the eastern branch of the other African apes. This extension takes them not only eastwards but also into more arid environments, away from the less seasonal tropical forests and into the savanna woodlands and grasslands. At the time in question – sometime after 7 million years ago – these were just taking their present shape. The process of rift formation was probably a part of this, along with a general deterioration in the climate. This emerging environment is one that is likely to have imposed entirely new selective pressures on these eastern apes, pressures that would have made a more terrestrial way of life beneficial – hence, perhaps, the evolution of bipedalism. Geographically, therefore, the distribution of the earliest fossils, while being a product of a geological accident, makes sense of the evolutionary patterns as well.[9]

There is perhaps something more to this than just contiguous geography. The Rift Valley is markedly different in terms of physiography and ecology from the rest of Africa. The constant tectonic upheavals, the eruptions of volcanoes and the deposition of great beds of lava and ash had the effect of cutting up the landscape into a series of relatively isolated lake basins. Throughout the last 10 million years this may well have been a most dynamic landscape. More importantly, perhaps, the lacustrine structure would not only have provided an interesting set of ecological circumstances for these exploratory bipedal apes, but would also have divided them up into isolated populations. It is a well known fact of evolutionary biology that new characteristics are more likely to be promoted and preserved in small populations. In large populations novelty will become swamped and lost, unable to overcome the genetic inertia. Small populations allow for novelty to survive. The fragmented environments of the eastern Rift Valley may thus have created evolutionary hotspots, not just for hominids, but also for many other species. Among the lakes and ashes diversity could flourish, and it is perhaps no coincidence that it is there that we find the earliest radical departure from ape morphology in the form of bipedalism, the greatest number of hominid forms, and, rather more speculatively, even later the emergence of modern humans.

Scepticism is always a healthy attitude to have towards the fossil record, and it is certainly necessary to be cautious about the view that the area that happens best to preserve fossils should also be the area that most promotes speciation. None the less it is tempting to draw the conclusion that the eastern parts of Africa were, for the hominids at least, not just a good place to die, but also a good place to evolve. Just as there are times when evolution seems to accelerate, so too there are places where evolutionary change and speciation is more likely to occur. The dynamism of the Rift Valley that promotes fossilization also seems to promote diversity of habitat and ultimately of evolutionary form.

The African Community

We can perhaps gain a firmer grasp of this by remembering that these hominids did not live nor evolve in isolation. It is not as if human ancestors evolved only when everything else had finished evolving. Far from it. At the time that hominids were so evolutionarily dynamic in Africa a host of other animals were also undergoing rapid evolutionary change. This is probably the best evidence we have that evolutionary events are not just randomly distributed through space and time, but that they occur in response to particular conditions and settings.

The African ecological community is the setting for hominid evolution.[10] At the time that the hominids were at their most diverse – that is, when there were the most species living at the same time – many other types of mammal were also radiating into different forms. Today there are just three types of pig in Africa (the warthog, the bushpig and the giant African forest pig). Two million years ago there may have been over twenty species. The lion and the hyaena are at present the only carnivores capable of hunting large prey. Two million years ago there were probably about ten species of large carnivore, some much larger than lions. The same pattern can be seen in a whole host of animals – antelopes, elephants, rhinos, as well as closer relatives such as the baboons.[11] Each of these groups reached maximum evolutionary diversity at about the same time.

There was nothing necessarily very African about all these groups – some, such as the lions, migrated into Africa from Asia. Others, though, were indigenous. What they all had in common was evolutionary opportunism. The climates of the tropical world were changing markedly from about 10 million years ago. In particular they were drying out, becoming more arid and more seasonal. The great expanses of equitable tropical rainforest that had previously spread like a great swathe across the whole continent were breaking up, fragmenting into smaller units, and being replaced by more open woodland and grassland. This shift in climate was an ecological disaster for many species – including most apes – to whom the rainforests were a natural home. But one species' disaster is another species' opportunity. For every species driven to extinction by these changes, another probably evolved in response to some new environmental situation. The hominids were nothing particularly special, and indeed Africa was nothing particularly special in evolutionary terms. It just happened that at the time when some apes were becoming more terrestrial the environments to which they were adapting were becoming more common. The range of African hominids reflected an evolutionary response, and one that was shared by a complete mammalian community. In one sense there was nothing terribly special about the African community other than that it happened to be the one in which hominids lived.

There is, though, another sense in which the African homeland is not so accidental. Obviously these climatic changes were occurring throughout the world. In the higher latitudes the decline in temperature that was to culminate in the ice ages was beginning, and these regions were moving from sub-tropical to temperate systems. Why was it not in these that hominids arose? There are two factors that are important in trying to answer this question.

The Constraints of the Past

The first of these factors relates to the nature of the evolutionary process itself. We think of natural selection as shaping organisms to fit their environment, and in this sense there is a notion of plants and animals being designed – albeit blindly – to suit the circumstances in which they live. It is all too easy to see this as similar to an engineer designing a machine, or a craftsman a piece of furniture. The end result in each case is something that works in a particular context.

There is, though, a fundamental difference. The engineer, in designing an engine will start from scratch, will construct the machinery from a plan in which there are no constraints other than the principles of engineering and the availability of resources. Evolutionary design is something very different. For a start there is no engineer, and, as we saw earlier, there is no fundamental end point, no intentionality. These are broad philosophical differences, but there is also a significant practical difference. Where the engineer can start from scratch and build up a new engine from entirely new pieces, evolutionary design can only use the existing material. It is as if you had to make a new aeroplane using only the bits that were used in making an existing aeroplane, with perhaps one or two new pieces. Change in this evolutionary context is much more constrained. The history of an organism is itself a major constraint on the future of that organism. Francois Jacob has referred to this as 'evolution the tinkerer'.[12] All evolutionary steps must consist of the modification of existing organs or behaviours. There can be no clearing away and starting again. In this sense evolution is in fact a very imperfect system, constantly cobbling together make-piece solutions, a Heath-Robinson affair rather than a sleek computer-aided design system.

When this general idea is applied to the particular problem at hand – why should hominids have arisen in Africa? – this has been considered solely in terms of why the conditions under which these remote ancestors lived in Africa were likely to lead to such evolutionary novelty. Certainly some of the answers are important, relating to patterns of environmental change and the diversity of environments. Equally important, though, is the nature of the organisms living there themselves. The animals that became human were not just a random selection of species faced with a particular environmental problem. Rather, it was the unique combination of a particular set of environmental problems and a particular type of animal. This can be seen most clearly by looking at the evolution of upright posture.

All hominids are characterized by bipedalism, the habitual upright posture and gait seen in humans. This is a major departure from the structure and behaviour of the other primates, and undoubtedly represents an adaptation to living an almost exclusively terrestrial, as opposed to arboreal, way of life. Indeed, it is probably the basis for all the things that make us human. It is also very easy to see how this fundamental trait links to the pattern of environmental change that occurred in Africa at the time that hominids were evolving. The forests were shrinking and the grasslands were expanding. This grassland was the place for any primate with an eye to survival, and the necessary precondition was a means of moving freely and efficiently around on the ground and away from the trees. 'Down from the trees' is an all too accurate slogan for human evolution.

This view underlines the basic idea that environments shape evolution. Where is the constraint of history in all this? The answer to this

The baboons represent the other group of primates that have extensively adapted to open and terrestrial environments.

question lies not in the hominids, but in other primate species. Hominids were not the only group of primates to face the problems of dwindling forests, nor the only ones to take up the opportunities offered by the new savannas. One group of monkeys in particular, the baboons, represents the other great line of primate savanna dwellers. Baboon is a very generalized term for a whole group of species that are endemic to large parts of Africa, that are primarily terrestrial and savanna-dwelling, and are, like their human counterparts, extremely successful. Few of the open habitats of Africa, including the margins of towns and agricultural areas, are not exploited by baboons. In the past they were even more diverse, and indeed they were one of the groups of animals that reached a peak of diversity about 2 million years ago, at the same time as the hominids. Their story is in many ways the same as that of our own lineage.

And yet they are not bipedal. Here is a primate species doing exactly the same as the hominids, colonizing new terrestrial environments, at the same time as the hominids, and yet pursuing an entirely different way of life. They are fully quadrupedal. Here seems to be the refutation of classic Darwinian adaptation – that environments shape organisms. Why should the same environment make hominid ancestors bipedal and baboons quadrupedal?

One possible answer is simply chance. After all, evolution is a random process in some ways. Perhaps the appropriate mutations had simply not arisen, and given time, they will. Such explanations are never terribly satisfactory and are certainly untestable. There is, though, an alternative, and that is the tight hand of history. An organism's evolution is shaped not just by its environment, but by what it already is, by its evolutionary past. Indeed, interaction between the past and the present is the real essence of evolution.

Baboons are monkeys, and are descended from monkeys that lived in trees, probably not dissimilar to the guenons of Africa today. Hominids are apes, and they are descended from other apes, that may have been similar to the living apes of Africa. Both lineages departed from their ancestral habitats to become more terrestrial. However, they had very different starting points and evolutionary histories. Although all were arboreal, their ancestors moved around the trees in very different ways. The ancestral monkeys were arboreal quadrupeds – they moved through the trees by running on all fours on top of the branches and then jumping quadrupedally from one tree or branch to another. The apes, though, tended to move around by suspending themselves below branches, or by clambering through the trees with their bodies held vertically. This difference was to have major consequences.[13]

When each group began to move to a more terrestrial life they carried with them their evolutionary baggage. The quickest route to a successful terrestrial life for a monkey was simply to adapt its quadrupedalism to the ground, to move in ways that we can see among so many ground-dwelling animals. The terrestrial apes, though, faced a very different problem. Already having a vertical posture and disproportionate lengths of arms and legs, quadrupedalism for them would have demanded a major reorientation of their entire behaviour and anatomy. For them, bipedalism was more likely to offer the shortest route to a successful evolutionary future on the ground.

The evolution of bipedalism has offered an insight into the evolutionary process. Evolution is not a wonderfully adjusted engineering system, but a short-term modifier of existing forms. Environments, through natural selection, play a major part in evolution, but they are not the only factor. The evolutionary past constrains and shapes the evolutionary future. Evolution is the result of the past and past adaptations interacting with current environments and selective pressures. Thus, when Africa is examined as the homeland of the human species, it is not just a question of the environments and the ecology of that continent. Equally important is its history, for this provides the biological raw material for subsequent evolutionary tinkering. In Europe or the Americas or Australia (which had no primates at all), or even Asia, the types of primate for which bipedalism was an appropriate adaptive solution to the problems of environmental change were simply not present. In Africa, they were. And this, of course, is why evolutionary questions tend to take one further and further back into the past, for with each answer there is a further question – for example, why did apes and monkeys develop such different postures in the first place. But answering that question is a temptation that must be resisted, for it will eventually end up in the primaeval swamp, whereas here it is necessary to pursue the human story forwards, not backwards.

History and the Laws of Evolution

I have answered the question of 'why Africa?' so far in terms of history – the history of apes, the history of their environments, and even the history of the interaction between the two. This gives a view of evolution in which the particulars of the past determine the particulars of the future. There is little here of the general laws of evolution that might be expected in a field of science. I said, though, that there are two possible answers, and the second, relating to the ecology of particular climatic regimes, does lead back to some more general principles.

When looking at the Rift Valley as a landscape in which species were likely to become isolated and therefore both to evolve novelty and to become extinct, the idea was introduced that some areas are more likely to promote speciation than others. In that case local differences within a region or across a continent were pertinent. There is, though, a bigger picture. If we look at the earth as a whole there is a distinct pattern to the distribution of species. It is broadly the case, whether for plants or animals at a number of taxonomic levels, that there are always more species near the equator than near the poles.[14] More accurately, there is a gradation from the tropics to the Arctic, with the number of species gradually declining. The tropics are more diverse than the mid-latitudes, which in turn are more diverse than the high latitudes. For example, land mammals range from 20 species at 66° to 160 at 8°. Orchids are perhaps the most dramatic gradient, with 2500 species at the equator, and only 15 at 66°.[15] This pattern runs across all forms of life and has implications for understanding the evolution of the human species.

Several explanations have been put forward to explain why this pattern should occur.[16] The most obvious of these relates to energy. The equator receives far more energy than the Arctic, and this creates far more opportunities for life. With more energy there is more plant life, which in turn creates further opportunities for herbivores, and where there are herbivores there are opportunities for carnivores. In the energy-rich environments of the tropics, complex ecosystems have evolved. This can easily be illustrated by the tropical rainforests of Africa and Amazonia, where the numbers of species in just a small area exceed those in the whole of the tundra and steppe regions. As most of these species have evolved locally, then it would be expected that there is far more evolutionary novelty in Africa and America and across parts of Asia than elsewhere.

This explanation is almost certainly broadly correct, although it can be refined in various ways to take into account other factors. For example, high latitudes have been much more subject to the ravages of climatic change, particularly the glaciations of the last few million years, and therefore they have suffered higher levels of extinction. The environments of the tropics have been more stable, and therefore new species have accumulated there. This in turn links in with the idea that diversity creates further diversity. Species respond to the availability of ecological niches, and indeed are classically thought to be defined by them. If there are many niches, then there are many species, but then in consequence these species become a niche for other species. A single tropical tree can become a niche for numerous ants, and so the increase of diversity becomes a self-perpetuating process.

Whatever the underlying reason, the relationship between evolutionary diversity and the tropics means that by and large most species would be expected to at least have their evolutionary origins close to the equator, even if they subsequently expand into other regions. For this reason alone it would be surprising for hominids not to have a tropical origin. It is interesting to reflect that in the discussions in the early part of the twentieth century about the probable locations of human origins these ecological considerations played little part. Darwin used phylogenetic relationships to argue for Africa, while others used the antiquity of civilization to support an Asian origin, while sheer nationalism probably underlay the preference for a French or British, or generally European one. Such considerations were unlikely to pinpoint Africa or indeed a part of Africa, none the less the essentially tropical distribution of primates and the evolutionary and ecological diversity of the tropics should have indicated that the warmer parts of the world were likely to be a more appropriate place to search for the origins of our lineage.

This amounts to a broad evolutionary rule, with general predictive power. Alone it cannot predict particular events, but when the question is asked about where hominid origins lie, the final answer turns out to encapsulate the nature of an evolutionary explanation. On the one hand it has elements of historical and geographical particularism – the distribution of existing species and the pattern of environmental change – but these are underpinned by a set of general patterns and principles that provide a basis for understanding these specific events and explaining their distribution in time and space. In this sense the search for the location of hominid origins provides as good an example as any of the nature of evolutionary biology. It is a science that does seek some general laws and principles, but the conditions under which these occur are themselves so often constrained by particular historical factors that they may be obscured. In the study of the human evolutionary story it is the particulars that have dominated, no doubt inspired by the unique nature of humans, but we can learn far more by returning at the end to the general principles, for they provide the framework for understanding a unique history.

Hominids as Outsiders

The African heritage in human evolution has probably been pursued as far as it legitimately can be, but it is possible to squeeze out one more insight by turning the whole question inside out. So far the focus has been on how large a role Africa plays, but perhaps the most obvious observation that anyone can make about humans and indeed many fossil hominids is the fact that they do not occur solely in Africa. Indeed, an excellent measure of human evolutionary success is the extent to which humans are found virtually everywhere today, and even the archaic species were widely distributed. This stands in stark contrast to the distribution of many species, which are particular to certain continents, regions and even localities.

By stressing Africa as the centre for evolutionary origins other regions have been relegated to the peripheries, and this implies that these were brought within the framework of hominid evolution by a process of colonization and dispersal. In most cases this is probably true. Hominids seem to have been particularly adept at expanding into and adapting to new environments, although the extent varies considerably through time.[17] For all of their early evolution hominids lived only in Africa, and probably only in the drier parts. From 1 million years ago they expanded across into Europe and Asia, gradually moving into more northerly regions. From at least 50,000 years ago they began the process of total global expansion, resulting in the explosive colonization of Australia, the Americas and most of the oceanic islands. Almost certainly this was not a single continuous event, and it was no doubt marked by numerous reversals and local extinctions, followed by recolonization, particularly in the earlier periods. None the less, the implication is that hominids were extremely successful intruders.

This pattern is very informative about the nature of hominids. They were clearly flexible and adaptable in their behaviour, tolerant of markedly different ecological and climatic conditions, and capable of demographic expansion. This is the classic ecological success story, although in terms of increased pressure on resources there was often a price to be paid. It also perhaps shows that dispersal into new regions, rather than increasing population density through intensification of foraging behaviour, was a hallmark of the hominids for most of their evolutionary history.[18] There is, though, another aspect that is intriguing.

Adaptation is thought of as natural selection shaping organisms to the requirements of their environment, and it follows that local species are best suited to the environments in which they evolved. In many people's minds is the classic notion of the perfect web of nature, with each species in its place, and performing its 'role' in the environment. Species from outside should be less well adapted, and when faced with competition from exogenous species we would perhaps expect the native to succeed at the expense of the newcomer.

There are many examples where this is true, and recent history is replete with stories of Europeans in particular, unsuccessfully trying to introduce their own crops and animals into alien environments, only to give up and adopt the local forms. American Thanksgiving is perhaps the only festival that celebrates this ecological process, for it is remembrance of the time when the first European colonists in the New World found that their Old World crops and stock had failed, and it was only the generosity of the indigenous American Indians in providing them with native turkeys and maize and squashes that allowed them to survive.

However, that same process of colonization is also characterized by remarkable and usually unintentional explosions of alien species in new environments.[19] Much of Africa and the Mediterranean is now overrun with eucalyptus trees, which are indigenous to Australia. Conversely the European rabbit spread rapidly and successfully throughout Australia. Horses in the New World, red deer in New Zealand, goats across many remote islands, are all evidence of this remarkable capacity of many species to thrive in alien environments without human help other than the actual introduction. In many cases this success is associated with subsequent enormous changes in the local habitats and the extinction of indigenous species.

To some extent these successful introductions can be explained in terms of the success of species that would always have done well, but had previously been unable to reach these new regions. In these cases the intruders were using environments in entirely new ways, and there were no local competitors capable of withstanding the introduction. The goats left on the islands of the southern hemisphere by sailors as a potential supply of meat on future voyages are an extreme example of this, and the success of placental mammals over marsupials in Australia may be another. But there are also examples, of which the grey squirrel in Europe is perhaps the best known, where there have not only been indigenous species exploiting the habitats, but also very closely related ones. Yet still the intruder has been successful.

Grey squirrels and eucalyptus trees seem a long way from hominid ancestors boldly going where none had gone before, but some connections can be made. It may well be that the distinction between areas where evolution has occurred and areas of dispersal is an important one, and not simply one of better or worse, or exciting and uninteresting from a fossil-hunting perspective. Different regions posed different problems for ancient populations, and the study of, say, the first modern humans of Europe is not made less interesting by the fact that they did not evolve there. Indeed, the very spectacular success that they appear to have had at the expense of the indigenous neanderthals may well have had its roots in the very fact that they *were* intruders. Like the rabbits in Australia, they were able to expand rapidly and to outcompete local populations. Long-term local adaptation and an intimate knowledge of and relationship with the surrounding environment may be no guarantee of success. Thus the different regions of the world, when viewed from the point of view of the palaeontologist, should not be seen as areas competing for the earliest fossils, but as varied landscapes on which two contrasting evolutionary events are repeatedly played out – the one, that of evolutionary change and innovation, the other, of population movement and expansion, migration and replacement, competition and extinction. As Mayr[20] pointed out many years ago, in evolution the centre and the edges are very different places to be. Indeed, many of the contrasting patterns in hominid evolution that we have seen in this and preceding chapters probably owe their origins to this very important distinction.

The conclusion to be drawn is that the fossil evidence reminds us that evolution does not really take place through time, it takes place in space, in particular places, and time is only the dimension in which we choose to measure it. In the specifics of hominid evolution each continent plays a different role, either a donor or a recipient. All are equally important in the final and overall picture. From the perspective of human evolution Africa lies at the centre, with each continent radiating from it. While there may often have been dispersals back into Africa, the general trend is out, to the north, the east, and the west.

African Eve – the Meaningless Mother

Africans can perhaps feel a certain amount of justifiable pride in their continent's role in hominid evolution, but as we saw at the beginning of this chapter, the idea of evolution as a race in which continents compete for pride of place is misleading, and as we have just discussed, indigenous and exogenous evolutionary events pose contrasting rather than ranked problems. In recent years, though, there has been a pronounced focus on Africa not just as the homeland of hominids, but as the place where our own species – *Homo sapiens* – was born.[21] Not only that, we have come to believe that the mother of all living humans was African. The African Eve, as the mother of the human race has become known, has struck a deep chord in a vision of the past, and because it is based on genetic evidence, it has acquired a force lacking in the rather haphazard nature of the fossil record. While at first sight this might seem the crowning glory of an African heritage, it is rather, an illusion that should not be allowed to obscure the African story told so far.

The African Eve relates not to the deep origins of the hominids some 5 million years ago, but to the origins of our own species, *Homo sapiens*, in the last 100,000 years or so. As was seen in chapter 4, again it was Africa that yielded the earliest fossil evidence. Modern humans – *Homo sapiens* – are really quite distinctive. They show a radical departure in skeletal design. Where the more archaic hominids have long, low crania (even with quite large brains), modern humans have high, rounded skulls. The faces of the archaics are big and often protuberant. Even the most massive-faced modern human is petite by comparison, with the face tucked neatly under the skull. Where the bones of the archaics are thick and robust, those of *Homo sapiens* are usually thinner and more slender. As a result, not only do they appear to be a relatively novel departure in evolutionary direction, they are also easily distinguished from other fossil hominids.

What is striking is the way they appear in the fossil record. In Europe, the continental home of the neanderthals, they appear suddenly around 40,000 to 30,000 years ago.[22] In South-east Asia and eastern Asia they appear at about the same time if not slightly later.[23] In Australia hominids make their appearance for the first time in characteristically modern form. In each case there seems to be no continuity between the native archaic populations and the modern forms. In contrast, in Africa modern *Homo sapiens* are found from certainly as far back as 110,000 years ago, and possibly as old as 140,000 years.[24] In other words, modern humans appear in Africa possibly as much as 100,000 years earlier than they do elsewhere in the world. African modern humans, in fact, seem to *pre-date* the neanderthals in Europe. Not only that, but where in Asia and Europe there is a general lack of transitional forms, in Africa, at sites such as Omo Kibbish in Ethiopia and Ngaloba in Tanzania, there are fossils that seem to show a mixture of modern and archaic features.[25]

This would seem to indicate that modern humans, like their distant forebears *Homo erectus* and like the earliest hominids themselves, have their origins in Africa, and possibly even in eastern Africa. Needless to say this simple story has its problems, not to say its critics.[26] It may be the case that there are links between archaic fossils from Java and the modern populations living in Australia. It is certainly the case that there were very old anatomically modern humans

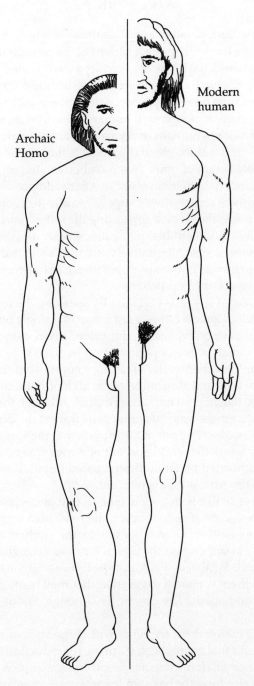

A comparison of an archaic and a modern *Homo*. Modern humans are more
gracile; the archaic forms tend to be much more robust, with heavy
musculature. However, the contrast is not complete as early
modern forms are quite heavily robust.

in the Middle East, possibly as old as those of Africa, and that they appear to have lived alongside a variant of the neanderthals. Clearly the hominid world 100,000 years ago was a complicated one. Indeed, where today we have only one species of hominid throughout the world, at that time there may have been as many as three.[27]

What happened at this point in human evolution? It seems that the evolution of modern humans in Africa resulted ultimately in the loss of the archaic forms throughout the world, although the process took a considerable length of time. Two possibilities suggest themselves. One is that modern humans evolved in Africa and possessed features which made them competitively superior to all other hominids. They thus swept across the globe, supplanting the indigenous inhabitants, who ignominiously dwindled into extinction. A replacement event, in evolutionary terms. Alternatively it could be that the spread of modern humans was not one of populations, but of genes, gradually swamping the archaic populations.

The fossil record suggests that broadly speaking the first of these is the more likely. However, this is not a case of a single point of origin and a single wave of expansion and colonization, but of continued evolutionary change over a period of up to 100,000 years, and successive or multiple dispersals.[28] This does not rule out the possibility of some gene flow and absorption of the archaic hominids of Europe and Asia into the modern human gene pool, although the most likely situation is that there was little or no gene flow at the extreme population contacts. Most important is the point that the spread of human populations across the world and out of Africa was neither a single event nor completed by an undifferentiated population. It is in this context that the African Eve is relevant.

The notion of an Eve is an ancient biblical one, and while the African Eve is an up-to-date genetic concept, the basic idea is the same. Eve is the ultimate mother, or, in other words, the mother of all humans living today. In some sense the search for human origins is, by definition, a search for Eve and Adam-type figures, although for any of these fossils there is no real suggestion that they really are ancestral to all later populations. The genetic Eve, though, is – or is supposed to be.

In all living matter there are genes within the nucleus of every cell. It is these genes that are replicated during reproduction and result in the inheritance of characteristics from one generation to another. These genes also form the basis for the way any organism grows and develops, and the bodily shape it takes on – its phenotype. These genes are known as nuclear genes. Outside the nucleus of the cell, though, lies a second set of genes, known as the mitochondria, or mitochondrial DNA (mtDNA). In contrast to the nuclear DNA, mtDNA plays no part in the development of the phenotype, but functions independently during the production of energy within the cell. In some ways mtDNA is best thought of as a benign and essential form of genetic parasite, living and reproducing independently within each cell.

Apart from its different function, mtDNA has a number of distinctive characteristics. First, compared with nuclear DNA there is not much of it – in technical terms, it is only 16,451 bases long, as opposed to the 30 million in nuclear genes. This makes it relatively easy to study and compare between individuals, populations and species. Secondly, and most important, it is inherited solely through the female line – that is, males do not pass on their mtDNA at all, and

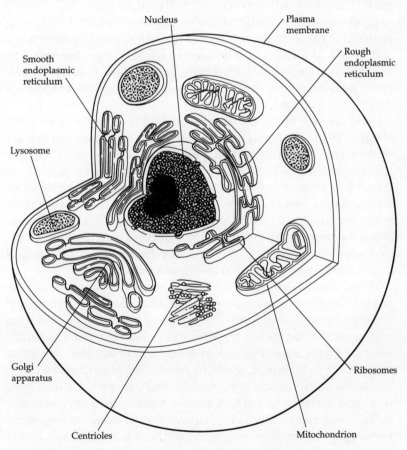

The structure of a cell. The nuclear DNA that forms the basis of development lies in the nucleus of the cell. The mitochondria lie outside the nucleus and exist as separate organelles playing a role in cell metabolism.

both males and females inherit their mtDNA from their mothers. Thirdly, parts of the mtDNA change relatively rapidly, making it an appropriate tool for investigating recent evolutionary history.[29]

The key point, though, is female inheritance. When tracing back nuclear genes, which may be inherited from either mothers or fathers, the path down which a gene is passed will constantly diverge. For example, an individual may inherit eye colour from either parent, and those parents in turn may have inherited it from either of their parents, and so on back; so that by the time the great-grandparents generation there are 16 potential ancestors for a particular trait. With mtDNA, though, this is not the case. Because inheritance is through the maternal line, then the pattern of inheritance must pass through a single lineage, and in turn all individuals' mtDNA ancestry will converge rather than diverge.

This characteristic of mtDNA makes it extremely useful for tracing evolutionary relationships. Variants of DNA, caused by mutations, and known as lineages, can be traced back to a single ancestor – an Eve. This is a simple and logical consequence of the pattern of inheritance, and it has been exploited by evolutionary geneticists to trace the evolution of modern humans.

There are two other important facts that need to be understood about mtDNA. The first is that the parts of the genetic system that are studied are apparently without function – that is, the particular sequences of mtDNA looked at serve no purpose, but are just 'sitting there'. This means that they are not subject to selection, which in turn means that the rate of change will simply be a function of the rate of mutation, which is usually thought to be constant. The second point is that it is possible, by looking at the sequences in detail, to work out how closely related to each other any two lineages or types of mtDNA are, and thus to build an evolutionary tree.

These various attributes have been employed to construct the genetic history of the human species. Wilson, Cann and Stoneking,[30] among others, have pioneered this type of analysis, and it is from their work that the idea of the African Eve has emerged. Essentially what they have done is to look at the pattern of variation in human mtDNA across the world, region by region, and population by population. What they found was twofold. First, it appeared that there was more variation in mtDNA among Africans than anywhere else in the world, and furthermore, that any 'tree' of relationships among human populations showed a major split between Africans and other populations. Secondly, using the estimates of the length of time it takes for change to occur within the mtDNA, they calculated that the last common ancestor of all the types of mtDNA found today lived approximately 200,000 years ago. As mtDNA is inherited maternally, this individual must have been a woman. As there is more variation in Africa than anywhere else, this implies that Eve's descendants have been in Africa longer than anywhere else in the world, and also that she must have lived in Africa. This is the basis for the African Eve.

There are a number of important implications in this work for the way the fossil record and the origins of modern humans are interpreted. Generally speaking the fossils indicate an African origin anyway, but the details are significant. Broadly speaking the mtDNA implies that of the alternatives outlined above – population movement or gene flow – the former is the more probable and consistent with the genetic data. Archaic hominids living elsewhere at this time could not have been descendants of Eve. More significantly, perhaps, the amount of genetic divergence suggests that humans are a very young species, as was indicated in chapter 4. All of this is perfectly reasonable, although there are a number of technical reasons for treating the results with a certain amount of caution.[31] Of concern here, though, is the nature of this African Eve, assuming that the geneticists are right.

It may seem that she reinforces the notion of an African heritage, and indeed personifies it. She was the first modern human, and from her we have inherited what makes us human. She was also African. However, in terms of the significant evolutionary and ecological processes that have been central to this chapter, she is remarkably unimportant. The principal reason for saying this is that, whatever else she was, she was not a modern human, and she was probably no different from thousands of females who lived before and after her. As discussed above, at between 300,000 and 200,000 years ago, she lived long before there is any evidence for anatomically modern humans, or any evidence for significant behavioural change. In the real terms of what it is to be a human as opposed to an archaic hominid, the African Eve tells us absolutely nothing. She is nothing more than a useful genetic marker for a lineage which subsequently underwent some major evolutionary changes, the consequences of which were undoubtedly extremely significant. She may also be a marker who indicates the fate of other populations by exclusion, and in that sense is informative about what happened in the past. However, we must remember the important limitations of this information, for we have already established that the mtDNA has, in this case, no function and no implications for the phenotype. The patterns of inheritance of the nuclear genes is far more complicated, and along with that goes the fact that the history of our species, as written in nuclear genes, is also far more complicated.

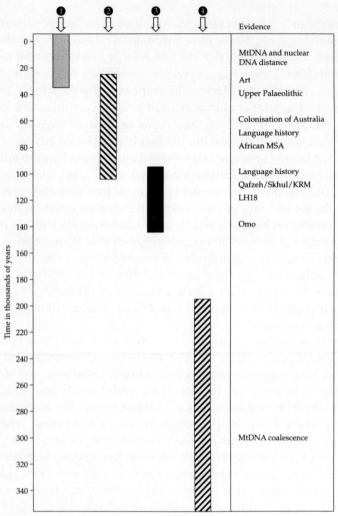

Evidence

MtDNA and nuclear
DNA distance

Art
Upper Palaeolithic

Colonisation of Australia
Language history
African MSA

Language history
Qafzeh/Skhul/KRM
LH18

Omo

MtDNA coalescence

Key
Range of estimates for:
❶ most modern population diversity
❷ the antiquity of modern behavioural capabilities
❸ the appearance of anatomically modern humans
❹ the last common (MtDNA) ancestor of living populations

Key events in the evolution of modern humans. Far from being a single event,
the evidence for the origin and diversification of modern humans is spread
over a considerable period of time. The mtDNA evidence relates to the
divergence of populations that are and are not ancestral to modern humans,
and this is likely to have occurred more than 200,000 years ago. The evidence
for the anatomical transition to modern humans is no older than 120,000 to
150,000 years ago. Evidence for behavioural changes that relate to full modern
human capabilities are seldom older than 50,000 years, while most of the
genetic evidence relating to human diversity suggests that considerable
changes occurred in the last 20,000 years.

The Mitochondrial Eve was a real mother – the nature of mtDNA inheritance means that this must be so. She must have been located somewhere and at some time. Although recent evidence suggests that the original interpretation may be more problematic, none the less the data does broadly point to her being in Africa, although the date could lie as far back as 300,000 years ago. The location of the African Eve can provide information about the broad scale of the evolution of *Homo sapiens*, and indeed the evidence of nuclear DNA is generally supportive of this notion,[32] but ironically the figure we are most certain existed – the mitochondrial mother – has perhaps less to indicate about the actual nature of the African heritage than all the other lines of evidence that have been explored in this chapter. The reason for this is that she cannot provide evidence for what hominids were doing in the past, let alone why they were doing it. For that, it is necessary to return to those humans before humanity, and ask why they looked the way they did. To understand why hominid evolution is largely a result of this African heritage, a very simple question must be asked – how did the characteristics that make a hominid a hominid give it an advantage in evolution?

7

Is Human Evolution Adaptive?

Thus far human evolution has been characterized in terms of passive populations bounced around by the vicissitudes of climate and geography. Species have appeared and disappeared with monotonous regularity, victims of a ruthless world of natural selection. It is as if hominid evolution consists of biological billiard balls – or perhaps snooker balls as there are so many of them – knocked around the table of evolution until they are pocketed into extinction. The analogy of a snooker ball is apt, because these species are characterless and regular, with smooth surfaces on which little can be discerned. There is no texture to evolution when described in this way.

There is some justification in this portrayal of the evolutionary past, because we have been looking at the overall pattern, the big picture. The lumps and bumps on the fossils have merely been markers of the passing of time and the divergence of species, not traits interesting in their own right. It has made very little difference how these species lived, and indeed it would probably have made very little difference so far whether the subject matter had been the evolution of humans or of humming birds. Perhaps that was one of the central points to be made – the pattern of human evolution is no different in outline from that of any group of animals.

And yet there are many very special things about humans, as there are in any species. In real evolution the snooker balls are not just different colours, but different shapes. And it is those different shapes that make them what they are. What is more, they are not just struck around the pool table of evolution by a *deus ex machina*, but play an active role in their fate. To go beyond the pattern of evolution it is necessary to look at the humans before humanity, and indeed humans themselves, not as anonymous species and populations, but as individuals, with patterns of behaviour that give them their unique character. In humans there are many strange features which need explaining – their upright posture, their intense sociality, their intelligence and their capacity for complex behaviour. These are the things that make us human, and it is these that need to be placed into an evolutionary framework. In this way it may be possible to move towards answering the major question posed at the beginning of this book – why are there human beings? Indeed the question can become more focused: given that humans represent something new and special in evolution, why do they take the particular form that they do? Why should it have been a naked bipedal African ape that moved into previously uncharted territories of evolution? This unique and curious stance and locomotion offers a way into the problem of why human evolution has taken the course it has.

Strategies of Survival

Species are the currency of evolution. It is species that come and go, and species that evolve. In the preceding chapters human evolution has been discussed in terms of species, and the pattern has been measured by looking at their origins and extinction. Evolutionary origins as a scientific problem is essentially a problem of species' origins. Speciation, extinction and diversity are the classic topics of evolutionary biology.

Despite the power and elegance of the species concept, though, there is something missing. When we think of humans, we do not think of the species as a whole, we think of the individual. It may well be a particular individual with particular characteristics that we wish to understand – why is he so aggressive, why am I so altruistic? – or it may be the individual as a symbol of humanity – man the political animal, or man the tropical animal. It is on individuals that the stamp of humanity lies, not on this amorphous concept of the species.

But the characteristics of a species are the characteristics of the individual members of that species, and its success is dependent upon those individuals. Evolutionary success is the sum of what its constituent parts add up to. As many evolutionary biologists have pointed out, despite the fact that evolution is a process played out over vast spans of time, none the less it is individuals that make up the unit of evolutionary bits. The reason for this is that it is individuals that reproduce, not species, and therefore it is on individuals that natural selection operates. After all, natural selection is about how many offspring an individual leaves behind – that is what is meant by differential reproductive success – so the key element must be the ability to reproduce. A species is evolutionarily successful when individuals within it have high rates of reproductive success, in relation both to each other and to other competitive species.

Looking at evolution in this way shows that species are not perfect spheres in evolutionary space, but complex masses of networks of reproducing individuals. The strategies they pursue to maximize their reproductive success keep these spheres rolling along without any external knocks. Natural selection changes the characteristics of the species by altering the individuals involved through the relative number of offspring they leave behind. Perhaps the most important implication of this is that evolution cannot be viewed simply as species passively responding to the changes in the environment around them. Rather, they are actively seeking new opportunities and solutions to problems that they face in their environment. This does not mean to imply that animals, let alone plants, are intentionally and self-consciously sitting there thinking up ways to succeed in evolution, but that they opportunistically act in ways that may or may not provide an evolutionary payoff. A tree, for example, may disperse its seeds far and wide into habitats that are currently beyond the range of its species distribution. Whether those seeds survive or not is a question of their adaptive suitability, but it is certainly not the case that oak trees, for example, are quietly sitting there contentedly accepting that they occur only on well drained soil or within a certain range of temperatures. If trees are actively pursuing evolutionary success, it is possible that intelligent, behaviourally flexible and ecologically tolerant species such as monkeys and apes may be doing the same thing.

We must start to think of evolution in terms of strategies.[1] Animals pursue strategies to acquire food, to avoid predators, to find mates and to produce offspring. The overall characteristics of these strategies are probably for most species characteristics shared by all the individuals as a whole. All woodpeckers are genetically programmed to drill into the bark of trees to extract insects. The success of particular individuals depends upon how well they deploy this general strategy – choosing the right trees at the right time of year, and correctly allocating the amount of time spent feeding. Differences in these particulars will determine which individuals are successful, which ones are not, success being measured in how far they enhance or decrease the number of offspring they leave behind. A woodpecker that spends its time pecking at a telegraph pole is unlikely to be a great evolutionary success. It is this deployment of strategies, for which some individuals will have more ability or more luck, that will drive the evolution of a species.

So when it comes to human evolution it is necessary to look at the strategies that the early hominids pursued, and how these shaped both basic biological characteristics – upright walking and large brains, for example – and behaviour. The appearance and disappearance of the species examined so far should be the outcome of the appropriateness of these strategies under different ecological and climatic conditions. Humans are what they are because the strategies of survival pursued by their particular ancestral hominids were successful, more successful than alternatives pursued by other individuals, and over time this came to shape and define the character of the hominids and of humans. Put simply, the characteristics that are 'human', both behaviourally and anatomically, are nothing more than the solutions preferred by natural selection out of the many strategies possible in the world of biological problems.

This chapter will try to use this very simple idea to show why humans have evolved their basic anatomical characteristics, characteristics that underlie much of human behaviour, before going on to pursue some of the more complex aspects of being human, such as the capacity for sociality and for complex thought and language.

Down from the Trees

No bar-room conversation is complete without reference to how recently someone has come down from the trees, or indeed if they really have. Tree dwelling is considered to be a true mark of evolutionary backwardness. In a book called the *The Evolution Man* by Roy Lewis,[2] the whole of human evolution is recounted in the trials and tribulations of one family, driven onward by the ambitious father's constant concern that he is falling behind in some evolutionary timetable. The true laggard in the family is Uncle Vanya, who is convinced that coming down from the trees is a fundamental mistake and still insists, despite the discomfort, in remaining arboreal. A terrestrial, ground-dwelling way of life seems to be an important part of being human.

And indeed, chapter 4 showed that while there are few things that link the whole range of hominids, one that does is bipedalism, and by association, terrestriality. Bipedalism is the fundamental characteristic that is found in all hominids. Contrary to popular belief, hominids became bipedal very much earlier than they became smart, and the whole of human anatomy, from the top of the skull to the tips of the toes, bears the marks of this way of upright walking

This major change in gait – and it contrasts very markedly with that of the apes – is usually thought of as an anatomical and essentially physical change. This is firmly in the realm of evolutionary biology. In a variable ape population those individuals that stand a little more upright have an advantage. Individuals with this genetic and anatomical endowment leave more offspring, and so over time the whole population and species becomes more upright in the way it moves, and so the evolution of bipedalism occurs. At this level, explaining bipedalism is probably neither very difficult nor particularly controversial or challenging in terms of evolutionary ideas. Yet, as was seen in the previous chapter, things are not that simple, and having the right ancestry was an important element in determining whether appropriate variations existed in the ancestral populations.

There is, though, another complication that makes the problem more interesting. While the presence of bipedalism can be observed and measured as a suite of anatomical characteristics, these are really better thought of as the necessary preconditions for the behaviours that they permit and facilitate. Bipedalism did not evolve because it was an aesthetically pleasing way of arranging a skeleton, but because it provided the functional basis for an individual – and remember that it was individuals that were bipedal – to carry out its daily life; in other words, bipedalism was part of a behavioural strategy. In evolution behavioural changes are more likely to precede major anatomical ones, and behaviour may be expected to provide the selective conditions under which novel anatomical configurations are likely to evolve.[3]

If evolution is about strategies of survival, and these strategies are ultimately behavioural, then bipedalism must be explained in term of the ways in which certain behaviours enhanced the reproduction of those early hominids possibly as much as 5 million years ago. That it later provided the basis for so many other things is neither here nor there. The question is, though, what reproductive advantage did this revolutionary form of walking give.

Speculation on this question is as old as the theory of evolution itself. Darwin saw it as the fundamental change, and argued that its advantages lay not as a means of gait, but as a by-product of other demands.[4] One of the great consequences of upright walking is that the hands are free to carry things. For most animals the forelimbs carry out the same basic function as the hindlimbs. They support the weight of the body and they propel it forward. Even for monkeys and apes this remains a major function. In their case, though, they also use their hands for grasping food and putting it in their mouths. The forelimbs and hands of these species are in fact an evolutionary compromise between these two demands. For bipedal hominids the need for this compromise is gone. The hands can be specialized simply for grasping. Darwin used this fact to argue that bipedalism was in a sense the price hominids had to pay for such dexterous hands, and that it was the behavioural requirements of tool-making and carrying that made it a part of hominid evolution.

There is still a lot to be said for this type of argument. Owen Lovejoy,[5] for one, has placed it into the context of a whole host of social and ecological connections. He has argued that the principal problem facing a large ape is that of infant survival. Large-bodied animals tend to reproduce slowly; for them a reproductive edge comes not from having many young, but from ensuring that as many as possible of the few they do have stand a chance of surviving to reproduce themselves. As these infants grow slowly their survival depends considerably on the extent to which their mothers can look after them. Parental care is the name of this particular evolutionary strategy. Lovejoy argues that the ability to carry food would greatly enhance the level of parental care. Males could provision females, releasing them from some of the burdens of foraging, giving them more time to spend looking after their young. Bipedalism would thus be a strategy that would enhance the reproductive success of both males and females.

Lovejoy's explanation illustrates perfectly the way in which major anatomical transformations are in fact rooted in behaviour, and indeed go far beyond just simple behaviours such as acquiring food and avoiding predators. The pattern of locomotion impinges upon the care of offspring and the whole social organization – for, as Lovejoy argues, for a male to provision a female and her young, as is the case with many birds, he must be sure that he is indeed the father of the young. If he is not then he is enhancing not his reproductive success but someone else's. Future generations would be populated not by noble, upright provisioning males, but by quadrupedal cheats.

In fact this lies at the heart of what is unsatisfactory about Lovejoy's model, alluring though it is. He sets out to explain not bipedalism, but monogamy. His model of the lives of the early hominids is one of nuclear happy families with the males off at work collecting food and the females staying at home to look after the young. It is certainly questionable whether this is what actually occurs in modern human populations, let alone in the distant past, and there are a number of reasons, as we shall see in the next chapter, why the social lives of extinct hominids were probably very different.

None the less, the rather dull anatomical features, described in the text books as the flaring of the ilium, the downward rotation of the foramen magnum, the development of lumbar curvature and the transverse and longitudinal arching of the foot – for that is what allows bipedalism – in fact relates more closely to the whole social and ecological fabric of the distant first hominids. The question still remains, though, why was it such a good survival strategy?

The Quest for Food

The period when the first hominids evolved was one of major climatic cooling. Globally the temperature was falling as the earth's climate drifted into the period known as the ice ages. The effect of the global decrease in temperatures in Africa was paradoxically to make the environment much drier. The lower temperatures meant that there was a lower level of atmospheric energy, and there was thus much less moisture carried around and available for rainfall. Over the eastern part of Africa the rainfall became reduced and the great bands of rainforest that had draped the whole region receded, to be replaced by more scattered woodland and even treeless grassland. This would have been a gradual and a patchy process, and it would be wrong to view this period – the Pliocene – as one of intense drought and desertification. For the most part there was more wood-land and richer savanna than is the case today.[6]

As was seen in the previous chapter, it was the expansion of these new environments that offered a set of opportunities to a whole range of animals and led to the evolution of the 'savanna community' of large mammals, including the hominids. The question to be answered is how did the hominids cope with this new environment and what problems did they face. Or rather, how did they grasp these new ecological opportunities, for it is not just a negative process of being forced out of an Eden-like homeland, but of populations being able to expand into new environments because they have found ways of tolerating new environments and conditions. This is not something special about humans, but simply the way evolution works.

What would have been the problems of living in these new environments?[7] For an arboreal primate the main difference would be that the trees were more widely spread and less suitable as a place both to live and to feed. In the forests apes such as chimpanzees are deeply attached to trees. They are sources of food and places to sleep and to escape from predators. The apes spend their days climbing up and down them, and their nights tucked up in arboreal nests. Fewer trees means more problems.

The first problem would be that more time would be spent on the ground moving between the trees. Instead of leaping from tree to tree or just scurrying a few metres from one to another, there would be distances of open ground to be covered. Moreover, the habitats would not have consisted of trees evenly spaced across the landscape. There would have been groves of trees separated by open stretches of grassland and bush.

There is another problem, and that is that the types of trees found in drier environments are very different from thick forest. Trees have to adapt just like animals, and where there is less rain the trees are more spindly, their leaves smaller and sometimes shed each year, and their fruit scarcer. Often the branches are set high on the trunk, so that terrestrial animals cannot reach them, and frequently thorns are profuse as a protection against predatory animals – that is, things that want to eat the leaves. These trees cannot have been easy to climb, and certainly would not have been very comfortable to spend much time in.

This is an environment where the resources that an ape would need – the fruits of trees and bushes – would have been much more

Open savanna habitats in Africa are thought to be the context in which much of early hominid evolution occurred. However, savanna habitats vary considerably in the extent to which the grass is mixed with isolated trees, bushes, or even semi-continuous woodland.

widely dispersed. It is not just that more time would have had to be spent on the ground, much greater distances would have had to be covered as well. The length of journey required each day to find enough food would have been much greater in these woodlands and savannas than in the forests. This can be seen today amongst those few populations of chimpanzees that live away from the forests – such as those found in the region of Mount Assirak in Senegal, studied by McGrew and Baldwin[8] – where the area over which the study community ranged was as much as three times larger than that of their forest counterparts.

These drier environments, with their dispersed trees, are the key to the very special features seen among the hominids, especially the bipedalism. To obtain the same amount of food as a chimpanzee in a forested environment, the hominids would have had to travel further. This in itself would have imposed higher energetic costs, and energy would therefore have been at a premium, even more so as the food itself may well have been scarce and often of poor nutritional quality. It is at this point that the first benefits of bipedalism would have become apparent. By and large quadrupedalism is energetically more efficient than bipedalism, certainly at any sort of higher speed. This is one reason why so many terrestrial animals are quadrupedal. But hominids did not evolve from typical quadrupedal animals; they evolved from semi-arboreal creatures whose own adaptations were a compromise between living in trees and living on the ground, involving considerable suspensory behaviour. A chimpanzee's pattern of knuckle-walking on the ground and a mixture of clambering and suspension in the trees is in fact an excellent compromise to a wide range of habitats. It is from this that bipedalism evolved, and it is with this, rather than true quadrupedalism, that energetic comparisons should be made. This reinforces the idea that the starting point in evolution is as important as the destination.

The hominids were spending less time in the trees and more on the ground, so the balance of compromise seen in chimpanzees is very different. Hominids require a more efficient means of moving across the ground, and compared with the way knuckle-walkers proceed, bipedalism is more efficient. Rodman and McHenry[9] have compared the energetics of chimpanzees and humans when walking – the former quadrupedally. They found that humans expend about one-third less energy when body size is taken into account. What this means is that a bipedal hominid could travel considerably further than a knuckle-walking chimpanzee for the same amount of energy. The furthest a male chimpanzee will travel in a day is about

11 kilometres; for the same level of energy expenditure, the bipedal hominid could go as far as 16 kilometres.[10]

In the context of less wooded environments this would be a major advantage, and would seem to account for the evolution of bipedalism without recourse to other attributes such as carrying, although these may well be further advantages that are permitted once bipedalism has evolved. This would certainly fit with the archaeological and palaeontological evidence, which seems to show that significant tool-making only appeared about 2 million years ago, whereas bipedalism may be as old as 4 or 5 million years. Furthermore, it does not necessarily presuppose that human ancestors were knuckle-walkers – for which there is very little evidence – but suggests that the antecedent pattern of locomotion was one that was less efficient than that of a typical quadruped such as a baboon or an antelope.

The selective pressures that made hominids upright were ecological ones – the problems of finding food in an environment where resources were widely dispersed and scarce. These pressures would have operated through behaviour. The actual process we would have seen had we been able to observe them would have been one of individual hominids travelling further and further to find sufficient food. Those individuals that managed to pursue these strategies of foraging over larger distances survived and reproduced better than those who did not; and those individuals that succeeded tended to be ones who were more upright, more bipedal.

Naked, Hot and Sweaty

Two things stand out when we look at the human mode of locomotion. The first is that it occurs in the earliest stages of evolution; the second, that it has demanded a major re-organization of the entire skeleton and muscular system. This stands as good evidence for its tremendous importance. Whatever led to this change, its consequences were major, and it is tempting to look for substantial causes as well. This is particularly the case when bipedalism is linked to two other very striking characteristics about this strange animal, *Homo sapiens*.

One is that humans do not have very much hair. Mammals are an inordinately hairy class of animals, from the shaggy yak to the svelte mink. The primates as a whole are no less hirsute. Humans stand alone, in Desmond Morris's striking phrase, as the naked ape.[11] Not quite alone throughout the mammals, as the elephant, the rhino, the hippo, the naked mole rat and a whole suite of marine mammals have also taken the evolutionary step of discarding their coats.

The other feature is that humans sweat copiously. All primates sweat, but when it comes to the density of sweat glands and the rate of sweating, there is no parallel. This is a major change, and one with a considerable cost, for it makes humans very thirsty creatures. Essentially humans use water to cool themselves, and as they evaporate so much water they have to replace it. This makes humans and hominids extremely dependent upon water and water sources.

Given these three very major changes – the three things that make people so visually and aromatically distinctive as animals – it is tempting to suggest that they may in some way be linked. Together they underpin the idea that this link is a major change in behaviour, and one that permeates all aspects of human life. While, unfortunately, pinpointing when in the past hair was lost is likely to be impossible, it is perhaps the whole suite of characters that needs explaining.

The key attribute is probably the loss of hair. Strictly speaking humans are not 'naked', but have extremely miniaturized hair over most of their bodies, although this varies dramatically with sex and geographical area. The zoologist Alistair Hardy[12] was one of the first to draw attention to the similarity with marine mammals, and Elaine Morgan[13] has done much to popularize the idea that humans became hairless because they went through an aquatic phase in the past. Both Hardy and Morgan interpreted these similarities in terms of environment: humans and marine mammals are hairless because they evolved in the water. Morgan in particular is also keen, quite rightly, to draw out the full behavioural context in which this evolutionary change could have occurred. Certainly for an ape to spend a few million years in the ocean is a dramatic evolutionary picture.[14]

While the aquatic-ape hypothesis was right to draw attention to the convergence of whales and humans, the conclusions it drew were wrong, not because it was an inappropriate comparison, but because it did not go far enough. The similarity is there because ancient hominids and whales did share a major problem, but the problem was not that of living in the water. The shared problem was the problem of controlling their temperature.

Marine mammals live in relatively cold environments. Their hairless bodies are an adaptation to streamlining, a convergence with the fish among which they swim. The trouble is that this reduces their level of insulation, and to compensate for this many of them have evolved thick layers of subcutaneous fat. These layers reduce the amount of heat loss.

As Peter Wheeler has shown, among the 'naked' animals that are terrestrial it is the opposite problem that is posed – that of keeping cool.[15] Very large animals are less able to cope with high ambient temperatures than small animals; they have to lose heat through their surface area, and in relation to the overall size of their bodies, this is small. Animals such as elephants have evolved very large ears as one way of increasing their surface area so as to be able to radiate heat into their environment. However, reducing the cover of hair over the body will also increase the radiating effect, and so elephants have only a thin, stubbly covering of hair. It is interesting to note here that the mammoth of the Pleistocene which lived in the very cold regions of Asia and Europe was hairy – as was the 'woolly rhino', whose tropical counterpart is also hairless.

The other and in many ways most striking of all the naked animals is the naked mole rat. This lives its entire life in burrows, and has some of the most bizarre and extreme social behaviour found among mammals. Regulating their temperature is a problem, and they use each other, and parts of the burrows, to absorb and disperse heat; to do this quickly a naked skin is an advantage. In this case it is probably not the actual temperature that is critical, but the very sensitive behavioural regulation of it, a fact that may have some interesting implications for looking at hairlessness.

It is easy to see how the sweating is also related to thermoregulation. Evaporating water secreted through glands creates a loss of heat in the immediate vicinity of the skin, and therefore cools the body by allowing more heat from the body's core to be dissipated to the skin. Wheeler, who has been the person responsible for bringing the whole question of thermoregulation to the forefront and has developed some extremely sophisticated quantitative models to demonstrate the significance of heat stress and its relationship to posture, has shown that a naked skin can more effectively radiate heat – and so nakedness and sweating are closely related to each other.[16]

What does this all mean for the hominid ancestors and their behavioural strategies of survival? Modern humans at least bear in their anatomy and physiology the marks of a species that has evolved in the context of extreme thermal stress. It is impossible to pinpoint when this stress may have occurred, but as bipedalism evolved very early, and may be related to thermal stress itself, it may well have been during the formative stages of the human lineage that something that the hominids were doing involved them getting hot. Walking across the landscape in search of food is critical both to the evolution of human thermal physiology and to human locomotor anatomy; it is also a critical behavioural strategy.

Space and Time

So far the problems the hominids would have had to face have been considered in terms of space – the strategies of movement they had to pursue to acquire food. It is the idea of a habitat in which food was much more patchily and thinly scattered that seems to make sense of the reasons why an upright ape would be at an advantage, why this should be a successful strategy of survival. To this must be added another element in order to understand the whole complex of bipedalism and the loss of body hair.

When resources are dispersed an animal is faced with a problem of distance and energy. However, as Robin Dunbar[17] has pointed out in the context of baboons, there is also a problem of time. To travel

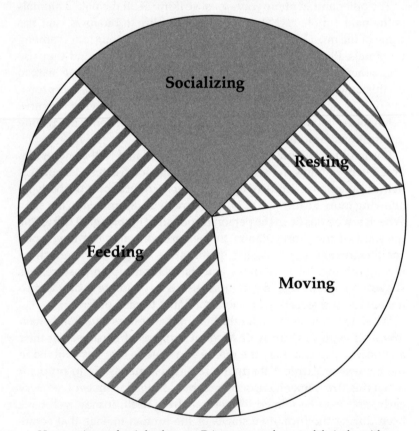

How a primate day is broken up. Primates spend most of their day either feeding or moving between food patches. The rest of the day is taken up in socializing and resting. As ecological stress occurs there will be an increase in the time needed for feeding and moving, and a loss of time available for social interactions or resting.

further takes longer, unless one travels faster. To travel faster is to expend more energy. There is thus a major cost involved. More importantly, while there may be more energy available, there is not necessarily more time. Time is a finite resource. For an animal that is active during the day – diurnal as opposed to nocturnal – the time available for foraging is limited by the hours of daylight. In equatorial regions this is fairly constant at about twelve hours a day. Thus if a hominid is forced to travel further for food, then there is less time for actual feeding. If the time spent feeding is kept constant, given that energy requirements must be satisfied, then there is less time for other activities, such as resting, or for social interactions, for getting on with the real job of evolution (reproducing), and caring for infants. Indeed, there is less time for sitting and thinking about how hard life is.

Bipedalism is therefore not just about space, and how to move through it, but also about time and how to utilize it. Dunbar has shown how this works among baboons. When resources are close together, then time spent travelling is reduced, and there is more time for feeding, resting and social interaction. As resources become more dispersed then time spent travelling goes up, and because time is finite, the time for social interactions and resting is reduced. Resting time is the first casualty of this, and then social time. But there is a limit to how far this can go.

It may well be advantageous to feed and live in large social groups, but the maintenance of these groups requires considerable time and effort. For individuals to get on well together demands that the relationships be lubricated and tested. This is the basis of social life. The larger the group, the more relationships there are to maintain. Yet if time for these social activities is constantly being eroded by increases in the time necessary for feeding, then the pressures are going to build up. Something will have to give. What Dunbar has shown is that in the end it is the group that goes. As baboon groups have to forage further and further their social groups come under more and more stress, induced by the pressures of time, and ultimately they are unable to maintain such large units – fission will occur, or some other mechanism will bring the group size down.[18]

The critical constraint, the actual factor determining the organization of the population, is time, or perhaps more specifically the relationship between the time available and the distribution of the resources required to sustain the population. How may this have applied to the hominids? As they moved into the more open habitats they would have been forced to forage further; bipedalism is an adaptation towards this that is energetically effective.[19] Another way of putting it is that they would have been time-stressed. One possibility is that, like the baboons, their social groups would have fragmented. That may certainly have been the case, but the physiological peculiarities of humans may point in another direction.

The more open environments led them to forage further and further afield, and their bipedalism enabled them to travel far enough to find sufficient food. This, though, took longer, and the quest for food would not only have eaten into their social and resting time, it would also have meant their either being active further into the hotter parts of the day or moving much more rapidly between food patches. In either case, heat would have been a problem. Walking fast or running raises the body temperature substantially, as all athletes know, while spending longer periods of time foraging would have meant being out and about as the temperature rose during the middle of the day. Solving the time stress would have led to heat stress. Here is the messiness of evolution; selection in one direction can lead directly to problems from another direction. By solving their energy problems through changes in both structure and behaviour hominids brought further problems – time stress affecting their social behaviour, and heat stress affecting their ability to carry out their activities.

Bipedalism may anatomically be a radical change, but its causes are not. It can be understood in terms of the incremental effects of apes foraging over longer distances. Small behavioural changes, though, can have major consequences. Anatomically that is certainly the case. Bipedalism, as Wheeler[20] has argued, may itself have helped reduce heat stress, by minimizing the area of the body that receives direct solar radiation. A quadruped receives direct heat from the sun across the whole of its head and back; by being upright this area is markedly reduced. It also removes more of the body from the reflected heat close to the ground. If this is correct, then bipedalism might itself be associated with the ability to increase the level of activity. Increased ability to sweat, to use evaporation off the skin, would have further helped keep the hominids cool, and this may have been one of the major changes at this time. Dependence upon sweating, though, has a major cost which flies yet again in the face of the need to forage further. Water lost through sweating must be replaced, and hominids are indeed highly water-dependent. Again, Peter Wheeler's models indicate the inter-relatedness of all these features, for bipedalism and loss of hair cover can reduce the level of heat load, and hence the use of water.

Such a mismatch, though, arises only because in thinking about the anatomy and physiology the behavioural strategies that underpin them have been forgotten. Hominids could become hot for two reasons. One was that they were in a hot environment; the other was that they overheated through activity. Best of all would have been to combine the two. It is the interaction of the behaviours of the hominids with the type of environment that is critical.

The key to loss of hair appears to be, then, not a question of heat or cold specifically, but the ability to regulate temperature relatively rapidly and efficiently. Broadly, human physiology shows the consequences for a species that has evolved under heat stress. What the details of relative hairlessness and copious sweating show is a lineage that can fine tune its temperature. That it is fine tuning rather than a broad adaptation to a single temperature is perhaps yet further evidence of the link between behaviour and anatomy, for it is really the fine tuning of daily activities that is underlying these patterns. Individuals and populations are balancing a whole host of often conflicting requirements. Essentially they relate to the manipulation of the time spent in different activities, in the context of a demanding environment – an environment in which food is widely scattered, temperatures are high, water is patchily distributed, and time for pursuing not just the strategies of survival but also the strategies of reproduction is limited.

A Day's Work in Evolution

Looking at evolution it is all too easy to be overwhelmed by the enormous lengths of time involved. In the previous chapter it was possible to show that placing greater emphasis on geography rather than chronology can expose new insights into the evolutionary process. In this chapter time has returned to the fore, but it is not the aeons of geological time, but the minutes and hours of daily activity.

The evolutionary events discussed here may have occurred over hundreds of thousands if not millions of years. That vast time scale, though, is not the scale at which evolutionary mechanisms work. If hominids became bipedal and naked they did so because these characteristics solved not the long-term problems, but the problems of day-to-day life. The countless individuals involved, generation after generation, in the process of evolution were merely trying to fit everything into a day on the savanna. To do so they had to forage further; to forage further they had to expend more energy, and bipedalism was a way of reducing that energy. But they still had to spend more time travelling, and this meant being active for greater lengths of time, and this in turn meant careful time budgeting in both the heat of the day and the coolness of the night. Hairlessness and sweating would have gone hand in hand with this time budgeting. The changes in the way time was spent during the course of a day had costs in terms of social interactions and the manageable size of groups.

At the outset of this book attention was drawn to the importance of time and place in considering human evolution; that the evolution of the hominids did not occur in a vacuum, but was a particular response to particular conditions, and that no understanding of human evolution was possible without knowing the details of time and place. In the past two chapters it is these that have been identified. The key lies in the stress caused to the behaviour of small pockets of apes in eastern Africa 5 or more million years ago as the habitats became more open, the climate drier, and the seasons more pronounced. Seasonal shortages, patchily and unpredictably distributed food, and the need to forage further and for longer were the triggers that led to new behaviours, and those behaviours in turn meant that the early hominid anatomy and physiology was exposed to novel selective pressures. It is the unique configuration of particular climatic conditions and particular evolutionary raw material influencing the daily lives of particular populations that underlies hominid origins, not the relentless passage of amorphous time.

In breaking up the shapeless species into the tangle of adaptive problems faced by individuals it has been possible to link together two time scales – the day-to-day life of an individual, with the aeons over which evolutionary change can occur. Perhaps more importantly, it has been shown how the very special features that characterize evolution's early hominids are not part of some massive detour into the world of the marine mammals, but are understandable as a series of knock-on consequences of adapting to a new environment. Solving one problem of survival inevitably seemed to lead to another one. Evolution the tinkerer was gradually tacking bits onto the hominid lineage as they grappled with their everyday lives.

But a problem remains. While the features of the earliest hominids – in particular their bipedalism – and their very distinctive characteristics as a primate and as a mammal are explicable in terms of the environments to which they are adapting, rather than as a grand sweep into monogamy, as Lovejoy has suggested, we are none the less left with robotic hominids. Efficient foragers, highly energetic animals sweeping across the plains of Africa while more lethargic primates fall behind in evolution, they seem a far cry from the socially tangled and cognitively complex humans that eventually evolved. Even Dunbar's baboons seem more realistic, for their ecological problems seem to be deeply interwoven with their social organization. In answering the question 'why are humans such strange animals?' in ecological terms, in terms of their strategies of survival, we have managed to lose sight of their other distinctive features – their sociality and their intelligence. Populations of any species, let alone early hominids, were not made up of undifferentiated, homogeneous individuals, but of males and females, young and old, large and small. Could it be that there is more to being human than being hot, naked, sweaty and overly energetic?

8

Why are Humans so Rare in Evolution?

Evolutionary Oddities

The evolutionary biologist J. B. S. Haldane once remarked that if he was to infer anything about God from a study of nature, it would be the creator's passion for beetles. What he had noted was one of the most fundamental aspects of ecology and evolution – different toxa vary enormously in their abundance and rarity, and while some groups are extremely diverse and varied, others are isolated and unique. Beetles occur in profusion, as do many other types of plants and animals. Others, on the other hand, stand as oddities, separated widely from other species. Among the mammals rodents are extremely common. Approximately 50 per cent of all living species of mammal are rodents. In contrast, the order Hyracoidea (a small African group, the hyraxes, which look like extremely large rats but are in fact more closely related to the elephant) has only a handful of species. In terms of numbers, large carnivores such as lions and hyaenas tend to be relatively rare, while small herbivores are at least an order of magnitude more common.

Explaining the ecological basis for this has been the mainstay of evolutionary ecology for many years. The trophic pyramid – that is, the way in which energy is lost as it moves up the food chain from plants to herbivores to carnivores – is the most obvious explanation for this pattern, while variation in the amount of energy in one part of the world compared with another helps explain why some regions are species-rich while others are impoverished. This indeed was part of the explanation for why Africa has been so important in the evolution of the primates and hominids.

This question of how common things are in evolution has mostly been directed at the broader issues of biogeography and community ecology. This perspective, though, can take us further in our pursuit of human evolution. However they are characterized, humans are essentially an extremely rare species – rare, that is, in the sense of closely similar species, rather than in terms of abundance or number of individuals. Humans are rare in that they are mammals, which, compared with other taxonomic groups such as insects or bacteria, are not really either very common or very important. They are rare in that they are social animals, and even rarer in being tool-makers, language-users and culture bearing. Indeed, once down to this level of specificity, they are unique.

This uniqueness poses a problem, or rather a paradox. We can arguably make the case that humans are the most successful species ever to have evolved. Such an argument cannot be sustained completely (after all, humans are a young species, as was seen in chapter 4, and so have not really stood the test of time), but it is possible to say that no other species has had such an impact on the earth, has been able to live so globally, has such a vast population, and has taken the limits of adaptation so far. While this may ultimately be the undoing of the species, at this stage it can be argued that humans have taken evolution to entirely new frontiers. Whatever it is that has made us human – bipedalism, brains, intelligence, culture and so on – has clearly given the species an enormous advantage and led to massive population expansion – the only objective measure of evolutionary success.

It is this that leads to the paradox. If human characteristics give such an advantage, why do all animals not have them. Surely selection would have favoured the appearance of these characteristics many times, possibly in widely divergent lineages. Surely language and culture, or simply generalized intelligence, should have emerged both early and rapidly in evolution, giving rise to many a science-fiction notion of talking hamsters or computers built by dolphins.

Mention of dolphins is of course an important reminder that the uniqueness of humans is to some extent an illusion or at least influenced by an anthropocentric notion of evolution. Other primates have evolved remarkable intelligence, as too have elephants and many of the marine mammals.[1] The other hominids that have been discussed throughout this book are clearly evidence for the same phenomena evolving several times in closely related species.

Tool-making has been found in a number of species, and of course there are many species capable of activities and behaviours that are far beyond humans. However, it would be only the most pedantic biologist who would argue that all this amounts to the same as the human evolutionary achievement. The paradox of human evolutionary success and its rarity allows the basic question to be posed in a novel way – what makes it so difficult to become human?

I shall try to pursue this question in the framework of recent work in comparative evolutionary biology, but before doing so it might be worth disposing of two very obvious answers to the questions about the fact that something like a human has only evolved once. The first is that evolution is essentially the product of chance, and therefore that it was just a matter of luck when 'something like a human' turned up, and equally that the hominids just happened to be the first one. This might be thought of as the mutationist view of evolution. The second is that this is something that could only happen once anyway, because the presence of 'something like humans' would inhibit the possibility of something else evolving in its place – a niche exclusion model.

The idea that evolution is simply a matter of chance is widely held. Evolution is often characterized as the random introduction of novelty, in technical terms through the occurrence of genetic mutations. Mutations are, for all intents and purposes, random; random, that is, in the sense that where, when and what sort of mutation will occur is not directed by any external parameter. In this way no organism can direct its own evolution by, for example, increasing the probability of a particularly favourable mutation occurring. This means that whether or not something new will turn up is indeed a matter of chance, and there is a strong random element in evolution. This has led to the view that evolution is itself a highly improbable process, most memorably likened to the chances of a whirlwind blowing through a workshop and putting together a Jumbo jet.

Emphasis on mutation has often been used by those antagonistic towards evolution as an argument against it. The classic example is in the case of the eye, discussed by Paley and virtually every critic of Darwin ever since. Given its complexity and the number of cells, nerves and different tissue types involved, it has been estimated that the probability of this set of co-adaptations evolving is infinitesimally small. Made up of millions of cells, interacting in different ways, and with a finely tuned function, it seems such an improbable product of evolution, if it is a random process, seemingly requiring more time to evolve than the age of the universe. The probability of humans evolving, with all their complex co-adaptations, is of course much

lower. It could therefore be argued that humans are, in the evolutionary context, extremely improbable. The question of why other
animals are not like humans does not arise, as such an improbable
event is unlikely to arise twice or more times. Indeed, the greater
miracle is that humans are here at all, and of course a creationist
would argue that the odds against it are so fantastically high that the
hand of the creator must lurk somewhere behind the scenes.

As Richard Dawkins has pointed out, though, this is a misunder-
standing of the role of chance in evolution.[2] Certainly chance is
important, but chance operates in conjunction with a highly deterministic mechanism, that of natural selection. The element of chance
in evolution occurs at various levels. The genetic is one, in that
mutations are errors in the process of replication, and, when they
occur, certainly have a strong stochastic element. At the population
level there are also chance events, such as where and when a volcano
may erupt or which populations are submerged under a rising sea
level. However, while the introduction of evolutionary novelty into
the world may be due to chance, its subsequent fate is certainly not.
Here natural selection and adaptation come into operation. Whether
a particular mutation will survive or not depends upon whether it
provides a survival and reproductive advantage. This is due not to
chance, but to competitive and environmental factors. Selection in fact
speeds up the rate of evolution quite considerably. Dawkins showed
this with an analogy, based on the old schoolchild's dream of revenge
against English teachers – that given enough time a monkey typing at
random could write the complete works of Shakespeare.

Dawkins took a single line from Hamlet – 'Methinks it is like a
weasel', and calculated that the ill-suited monkey would have a 1 in
10^3 chance of getting just this line right. Allowing only minimal time
for the poor monkey and its descendants to continue eating and
reproducing in a remarkably benign and constant environment, and
assuming a reasonable typing speed, this would take nearly a billion,
billion, billion years. However, Dawkins introduced a form of natural selection into these random variations: each time a letter was in
the right place (i.e., was the same as in the quotation) it became fixed.
In Darwinian terms it was fitter and could not be displaced by alternative letters. When selection operated, the goal could be achieved in
the lifespan of a single monkey, and indeed within, on average, fifty
trial attempts.

When applied to the problem of the probability of humans evolving, natural selection brings back into the framework the question of
other animals. If natural selection gives such an advantage to
humans and human characteristics, then perhaps not only should it

have recurred in several lineages, but one might even ask the question, why not earlier? The answer to why not earlier might well be provided by reminding ourselves, as was discovered in chapter 6, that the appearance of new characteristics is dependent upon historical circumstances and prior adaptations, and these may only have occurred in the context of pre-existing basal hominid features. However, the widespread trends towards larger brains, towards sociality in primates and other mammals, and towards other signs of increased behavioural flexibility in larger mammals may be taken as evidence that humans are part of a selective continuum. It can be inferred that there are both strong and widespread selective pressures in favour of the sort of behavioural and biological characteristics found in humans, and also constraints against them evolving. The paradox thus survives.

The second solution to the problem is that of arguing that once humans were around, then their niche was so overwhelming and all-embracing that no other species could ever evolve in the direction of humans. This argument has in fact been used in a number of ways. Wolpoff employed it to argue against the existence of more than one species of hominid in the early African Pleistocene.[3] All hominids are culture-bearing, he suggested, and as culture allows hominids to occupy all habitats, therefore there can only ever be one hominid. It is known from the fossil record that this theory simply does not hold, and the evidence shows that for much of the time more than one hominid species did exist, and indeed that some hominid character-istics may have evolved more than once and in parallel. The adaptive radiations that have been explored are further evidence for the selec-tive advantages of broadly human characteristics, and their widespread occurrence among the hominids.

Even more startling has been Kortlandt's dehumanization hypoth-esis.[4] He has taken the all-embracing nature of human adaptation even further and argued that over millions of years of co-existence in African hominids and humans have acted as a severe selective pres-sure on chimpanzees and have led them to become 'less human' in order to avoid competition with humans. The last common ancestor of chimpanzees and humans, according to this model, would be more human-like than modern chimpanzees.

Attractive as this model is, there is little evidence to support it in its strong form. In a weaker form, Kortlandt has certainly made an important ecological argument, but again one that is to some extent contradicted by the co-existence of a number of hominid species for much of their evolutionary history.

While a case may be made that modern humans are unlikely to

permit the survival of very closely related taxa, and that the demise of the archaic hominids over the last few hundred thousand years is evidence of this intolerance, this argument cannot apply more generally to the overall sweep of hominid evolution. To explain the rarity of the hominid if not the human phenomenon, we must look not only at the benefits of being human, but the costs as well.

The Price of Evolutionary Success

It is only natural that when humans compare themselves with other animals they tend to think in terms of the advantages they have over them, and to think of the various traits that are human as enormously beneficial. If people do think in terms of the disadvantages of humans, then they focus either on a few anatomical discomforts – lower back pain, the difficulty of childbirth – or else on the suggestion that humans suffer from various evolutionary excesses. Human brains, it can be argued, have become too powerful, and so humans are now out of control as a species. For the most part, though, humans can smugly look down on most species and consider our adaptations to be, on balance, beneficial.

However, the question of evolutionary rarity imposes a new question. If human features are so beneficial then they should have evolved many times, and if they have not, then it must be because there are severe costs or disadvantages involved. This is the ruthless logic of the evolutionary process.

One way of looking at evolution is in terms of costs and benefits.[5] New features arise because they are advantageous, they provide some benefit. That benefit is usually both direct and indirect. The direct benefit might be strictly functional. For example, the very strange Madagascan primate the aye-aye (*Daubentonia*) has a markedly extended middle finger on its hands. This is known as a needle-claw, being long and thin. It is used for poking into and behind the bark of trees, and extracting insects to eat. Without this claw-like finger the aye-aye would not be able to feed as efficiently, and so it is possible to say with some confidence that the evolution of longer and longer fingers provided direct benefits. That it evolved at all, though, was because it also provided indirect benefits, that of increased reproductive success. Those individuals with relatively longer fingers had an energetic advantage that translated into more offspring, and hence more descendants than their shorter-fingered counterparts. The ecological or energetic advantages led to reproductive benefits, and hence evolutionary success.

New characteristics therefore evolve and spread because of the benefits involved. This, though, is only half the evolutionary equation. No biological organ just provides benefits; it comes with costs as well. Two types of costs are immediately obvious. The first is a metabolic and energetic one. A longer finger has more cells, will take longer to grow, and is therefore 'more expensive' than a short finger. In considering the evolution of the needle-claw on the aye-aye we want to place the benefits – increased efficiency in gaining energy from insects hidden behind the bark – against the costs of growing and maintaining a larger finger and all the associated muscles and neural control. In this case, these costs might be negligible, but it is easy to think of examples where the costs are far from negligible. Horses have increased very markedly in size since the first ones evolved in the Eocene, from around 20 kilograms to the 400 kilograms of the modern zebra.[6] A number of explanations for this increase in size might be suggested. One is that a large animal is less subject to predators than a smaller animal. By becoming larger the horse lineage received the benefit of release from smaller predators (although, of course, it paid the price of becoming more attractive to large predators). More importantly, though, the horses paid the considerably greater energetic costs of more than doubling body size, a change that must have entirely transformed the biology of the group. Once again, though, the benefits exceeded the costs.

The second type of costs are what economists refer to as opportunity costs. When someone decides to spend money on some new consumer good or to invest in a company, they are not just imposing a set of direct costs related to the price involved. By investing money in one product, then a cost is also paid through the loss of opportunity to use the resources for something else. Furthermore, opportunity costs are not simply measurable in terms of other spending options, but are paid also in terms of the time spent doing one thing rather than another, or being in the wrong place and therefore missing other advantageous options.

Evolutionary change also imposes opportunity costs. The long neck of the giraffe is extremely beneficial for foraging in tree tops, but there are clearly opportunity costs when it comes to being able to move around in a dense tropical forest. A lion deciding to chase a small Thomson's gazelle may pay an opportunity cost when there is a much larger and tastier wildebeest a few hundred metres further along. Each step taken down one evolutionary pathway closes off several others.[7]

This is a very similar argument to that made in chapter 6 concerning the importance of historical factors in determining evolutionary trajectories; past events impose costs on all organisms in terms of their evolutionary potential, and all evolutionary adaptations also impose direct costs relating to the growth and maintenance of the physical equipment necessary to carry them out. How does this help to understand human evolution?

The Costs and Benefits of Being Human

When we move from the generalities of evolutionary theory to the specifics of human evolution, it is possible to turn the cost/benefit model towards answering the questions about the reasons for the rarity of the human species. If humans as an evolutionary class are rare, the thinking about the costs involved in 'being human' opens up two possible, albeit related, answers.

The first answer is that the selective pressures were not strong enough. If novel features evolve when the benefits of those features provide a reproductive advantage, then it is obvious that human features can be expected to have allowed those individuals who were 'more hominid' or 'more human' to have a higher level of reproductive success, and to leave more surviving offspring. Selection being the term given to the differences in reproductive success between different morphs or strategies, if there had been insufficient advantageous differences in reproductive success between the more human and the less human, then the features such as bipedalism or larger brains would not have evolved. The inevitable conclusion would be that being human provides few evolutionary or adaptive advantages that are favoured by selection – that the selective environment 'prefers' the run-of-the-mill mammals and apes.

This first answer, of course, hides the second answer. It is clear that there are advantages in being human, and it is easy enough to posit any number of benefits that might accrue from being bipedal (reaching higher fruits or making tools) or being intelligent (planning ahead and scheming the way up the social ladder) or being hairless (evaporating water more easily, and hence remaining cooler). What is hidden is the reason why the benefits might not have been sufficient, and therefore why the selective pressures were not strong enough. For each of these features there are costs – in being bipedal, the costs of no longer being able to climb trees as easily; in being naked, exposure to harmful radiation; in being intelligent, endless indecision, perhaps. It is not that the features do not have benefits, but that these benefits do not, for most species, exceed the costs involved.

This allows the realization that the question about the rarity or otherwise of humans and other taxa is in fact another question in disguise, a much more technical and answerable question: under what conditions do the benefits of possessing hominid or human characteristics exceed the costs. An answer to this question is potentially very significant. First, it provides an answer to why there is such a thing as humanity within the biological world, a not uninteresting problem. Secondly, posed in this way it gets around the anthropocentrism and circularity of much anthropology because it is phrased in such a way as simultaneously to address the problem of both the presence and the absence (or abundance and rarity) of different evolutionary lineages. And thirdly, it places the focus on not only the evolving species of hominids, but the contexts and environments in which they evolved. This leads beyond just the problem of human existence, and focuses on the specifics of the time and place in which it happened.

Brains and the Importance of Size

Having skirmished around the outer defences of human uniqueness, the time has perhaps come to take on the more difficult question of producing an evolutionary explanation for what really makes humans both different and rare within the realms of evolutionary biology. Bipedalism may be important and interesting, but it is, after all, just a way of getting around, and one shared with birds. Being naked and sweaty may be striking among the mammals, but it can hardly be the most important element of the modern world, especially as many people spend most of the time trying to be less naked and less sweaty. What is important, though, is that humans are an extremely clever species, with an enormous capacity to think, talk, understand, and behave in ways that have ramifications far beyond the individual. Humans also live in social groups, and survivorship, as noted in chapter 3, depends almost entirely upon the ability to maintain social relationships.

Chapter 3 also looked at the problem of the gap between humans and other animals alive today. While this gap may be an illusion, in that there are a variety of hominids known from the fossil record which are the intermediaries between the human and animal worlds, none the less, the problem remains: just how different are humans when it comes to intelligence? Comparing the cognitive skills of different species is fraught with problems. Some approaches tend to enhance the differences. After all, it is we humans that are doing the testing, and so there must be an inbuilt bias towards skills which are dependent upon the human context. It is perhaps no accident that so much research has focused on language and communication skills, for human intelligence is very closely linked to language. This may not be the case in other species. Rodents, for example, would certainly score poorly when it comes to language-based cognition, but on the other hand have an extraordinary capacity for solving three-dimensional spatial problems.

On the other hand there are also assumptions that tend to downgrade animal intelligence. Much early research was fraught with the problem of anthropomorphizing animal abilities. There is a natural tendency to want the animals that are the subject of experiments or observations to do well, and indeed the whole point of the work is often to show how clever a particular primate is. People who do not think chimpanzees are capable of language do not often choose to work in this field. Historically therefore many skills have been imputed without sufficient unambiguous evidence. To counterbalance this there is now a tendency not to make any assumptions or draw any conclusions unless the evidence is absolutely watertight. This means that in most studies these days most animals are assumed to be unintelligent, or lacking in consciousness, or incapable of a particular skill, unless there is complete proof to the contrary. Other, more parsimonious explanations are preferred, and the burden of proof almost lies with the animal. The result is that many psychologists today are probably underestimating the skills of their subject animals in order to avoid the pitfalls of anthropomorphism.

Given these problems, what can we say about human intelligence in a comparative perspective? Focusing specifically on close primate relatives, there is general consensus that chimpanzees do not have language in the sense that modern humans do. Beyond that, consensus falls away, although certain baselines can be established. Recognition of individuals, rather than categories or just conspecifics, is widespread among mammals and birds. Identifying different categories of predator has been seen in vervet monkeys, and other monkeys as well as other species are capable of assessing the quality and abundance of food sources. Lions and gibbons have been shown to classify individuals into strangers and known individuals, and also to be able to count individuals. Elephants and most primates are very astute at assigning and assessing social rank and sexual condition to other individuals. Social groupings are usually based on kinship, and most social species recognize kin from non-kin to quite a sophisticated degree. Beyond knowing their own relationship to other individuals, members of many primate species have been shown to know and classify third-party relationships, those between two other individuals. In addition, there is extensive evidence in many species for detailed knowledge of the environment, for the ability to navigate, and to carry out complex manipulation of elements of the environment, up to and including making tools. And for chimpanzees at least, their capabilities stretch to manipulating relationships, deception of other individuals, and empathy with another's emotional state.[8]

While these conclusions drawn from other extant species provide a basic framework, they suffer from a serious drawback. Few of these capabilities could be inferred with confidence from the fossil record or with archaeological techniques, and they thus relate little to the many hominid species that lie in the evolutionary space in which modern humans actually evolved. We are thus thrown back a stage, and need to find something that can be compared across species and across time. The size of the brain seems an obvious solution.

Humans have an average brain size of around 1400 cubic centimetres, or 1400 grams. Chimpanzees' brains are on average 400 grams, or about one quarter of the size. An average Old World monkey such as a small baboon will have a brain size of approximately 100 grams. This in turn can be compared with an equivalent-sized mammal such as a dog or a moderate-sized antelope, whose brain would be less than 50 grams. Furthermore, the various fossil hominids that have been discussed can also be plotted on this line of increasing and decreasing brain size – the first

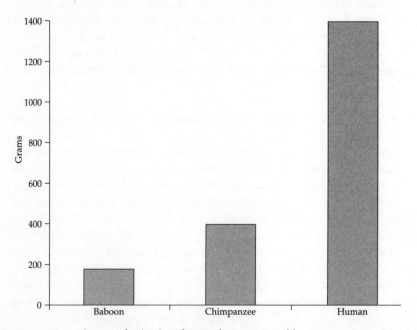

Average brain sizes for monkeys, apes and humans.

australopithecines, at around 450 grams; the first *Homo,* 2 million years ago, around 700 grams; and *Homo erectus,* between 800 and 1200 grams.

Brain size thus seems an ideal surrogate for measuring intelligence. Indeed, it is hardly a surrogate, as the brain is itself the organ which carries out the information-processing and thinking. The size of the human brain, being so much larger than any close relative, seems to be confirmatory evidence for the extent to which humans are significantly different from, and also more intelligent than, any other species. When brain size is plotted against time we can furthermore see both the inexorable rise of brains during the course of geological time, but also the relatively rapid changes that have occurred more specifically during later hominid evolution. Humans are rare not just in their behaviour, but in the anatomical and neurobiological foundations of that behaviour.

There is, of course, a hitch in this argument. Looking more broadly at the size of brains shows that the human brain is not the largest. The elephant has a brain size of between 4.5 and 5 kilograms, considerably larger than that of humans. The brains of the largest whales may

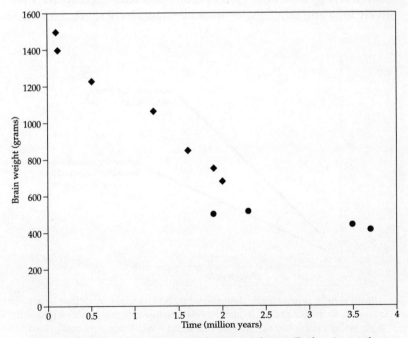

The pattern of brain size increase in human evolution. Each point on the graph represents the brain size of a hominid taxa at the time of its first appearance in the fossil record. Circles are australopithecines and diamonds are *Homo.*

go up to over 10 kilograms, almost ten times the size of a human brain. Should one therefore conclude that humans are not that rare, and that many other species have evolved the same capacities or the same potential at least? Perhaps, but to do so without first considering the effect of size in general would be premature.

At first sight it might seem simplest to use the relative brain size rather than absolute size. Human brains are approximately 3 per cent of body weight, whereas elephant brains are only around 0.2 per cent. Chimpanzees' brains would be around 1 per cent. The natural order is restored. Unfortunately, this approach can be misleading, as it assumes that all parts of the body will increase in size at the same rate (isometry), and that size itself does not have an effect on other aspects of biology.

There has been a long history of research into the understanding of the relationship between size and shape in biology. Such pioneers as D'Arcy Thomson and Julian Huxley[9] established in the early part of the twentieth century that there are certain underlying principles and generalities in the way in which the size and shape of individual organs will change as the size of an animal increases. This can be seen most clearly by looking at the legs of animals. Small gazelles have relatively long and thin legs, whereas elephants have short, thick ones.

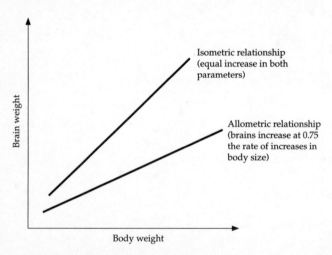

Brain weight is principally determined by body weight, such that larger animals have larger brains. In fact the relationship between brain weight and body weight is allometric – that is, that brain weight increases more slowly than body weight. Larger animals therefore have relatively small brains, even though they are absolutely larger. The precise nature of the allometric relationship makes it possible to draw significant inferences about the evolution of brains in primate and hominid evolution.

Relatively speaking one might expect them to be equally long or equally thick. In fact the diameter of a leg increases much more rapidly with body size than its length. The reason is that the legs have to bear the weight of the animal, and as this is a volumetric measure it is increasing much faster than height; increases in height are in fact only the cube root of the increases in weight.

This relationship between shape and size is known as allometry. The word expresses the basic idea that different things grow at different rates. Allometric relationships can be explored by plotting on a graph the overall size of an animal and the size of its constituent parts – in this case, of course, the brain. If both increase at the same rate, then there is isometry (the slope of the line on the graph will be 45°). If the rates are different, then there is an allometric factor to be taken into account when calculating relative size. This can be done by calculating an equation to predict the size of the brain that would be expected for a particular size of animal. For example, the equation for predicting brain size from body size in mammals is:

expected brain size = (0.76 x body weight) + 1.77

(for this equation it is necessary to transform the values of both body weight and brain weight into their logarithms.)

For an elephant this might be as follows:

expected brain size = (0.76 x log 2,766,000 grams body weight) + 1.77

or

expected brain size = 4,632 grams or 4.6 kilograms

If the expected and the actual or observed brain sizes are compared, they are very similar – around 4.75 kilograms. From this it can be concluded that an elephant has about the right size of brain for its body size. This can be formalized by calculating what is known as an encephalization quotient or EQ,[10] which is a measure of the relationship between actual and expected brain size, and is calculated simply by dividing the former by the latter:

EQ = observed brain size / expected brain size

or in the case of the elephant:

EQ = 4.75 / 4.6 = 1.03

An EQ of 1 means that the observed and expected sizes are about the same (a typical mammal). Greater than 1 means that the animal has a larger brain than expected, while less than 1 is below average. In fact this approach shows that the simple use of percentages would have been misleading because as animals become larger their brains do increase in size, but not as fast as their body size. Large animals, other things being equal, have relatively small brains.

There has been a purpose in going through the mathematics of this in some detail as we shall be returning to it later, when the numbers turn out to yield still more information. What we have for the moment, though, is what should be a more meaningful way of comparing brain size across time. Calculating EQs shows that humans have EQs of around 7.0, chimpanzees around 2.4, gorillas about 1.14. More significantly, the primates as a whole all have EQs greater than 1, showing that primate evolution has generally been characterized by the evolution of larger brains.

Calculating EQs for the extinct hominids is not quite as simple, as it requires being able to estimate not just the brain size but also the body size.[11] This means deriving similar equations that predict overall body mass from particular measurements on individual bones

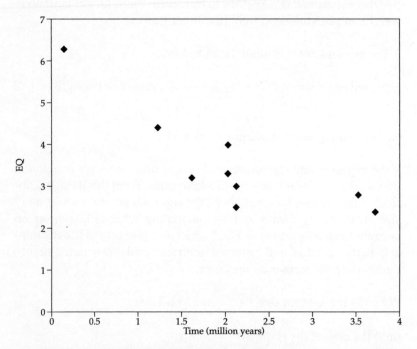

Relative brain size, measured here as EQ or encephalization quotient, of fossil hominid taxa shown in relation to time.

(for example, the cross-sectional area of the femur). When these calculation are made, the pattern of change in EQ of the hominids is not constant. The early hominids – the australopithecines – have encephalization quotients of between 2.2 and 2.9. These are not unlike that of the chimpanzee. Early *Homo* (up to about 1.0 million years ago) had an EQ of up to 3.5, and later *Homo* exceeded 4.0. Over the last 300,000 years or so there was a relatively rapid increase in EQ, up to the modern human figure of between 6.0 and 7.0.[12]

EQs enable the comparison of species of different sizes and also across time. EQs do demonstrate that humans have brains that are in size as unique as the behaviours which they generate. Primates appear to be generally large-brained and confirm an intuitive understanding of them as intelligent animals. However, the closest in EQ to humans are in fact not the apes, but the dolphins.[13]

Clearly there is more to intelligence than just the size of the brain. There is some evidence that in the evolution of intelligence not all parts of the brain have evolved equally, and not all are equally important. The neocortex and the pre-frontal lobe may be particularly crucial to the kinds of intelligence that are considered interesting in humans and other primates, and measurements of these can be useful. However, at the very general scale used here, they show much the same sorts of result. It is also important to emphasize that in looking at the grand scheme of brain evolution we are discussing changes in brain size over several orders of magnitude, and even within the hominids there is at least a doubling of size. This scale of variation is much greater than that found in humans, and there is little or no evidence to suggest that such small-scale variation is particularly significant within the human species.

Examining the comparative nature of brains and intelligence in humans, monkeys and mammals has shown that while what is seen in humans can be placed on a continuum with other species, especially if fossil hominids are taken into account, it remains true that species that approach the human condition are, in the overall scale of nature, relatively rare. What we must ask, therefore, is to what other factors in evolution can intelligence be related. In other words, why be smart?

Why be Smart?

Anyone with a knowledge of animals can see that some species are clearly very smart, others are very dumb. The animals selected for circuses and zoo entertainment tend to be the ones that respond well to human teaching, and these are among the smartest – horses, seals, elephants, chimpanzees and dolphins. In contrast, to watch a buffalo or a wildebeest standing morosely and inactively in the African savanna does not inspire the thought that one is in the presence of a zoological genius. According to the costs and benefits model of evolution that has been developed in this chapter the smarter animals have evolved their large brains because they derive benefits from it. What, then, are the benefits of being intelligent – what is intelligence *for*?

Two basic answers have been provided to this question. The first is what may be described as the ecologists' answer.[14] Some animals have larger brains than others because they require this greater intelligence to be able to survive in their environment. A number of different factors may be involved here. Predators, for example, depend for their food upon prey items, which may move very fast, may hide, may take evasive action, or may live in different areas at different times. Their environment is therefore more complicated than that of, say, a grass-eater living on a lawn. Responding to environmental complexity and change should therefore demand a larger brain for greater levels of information-processing. This holds for carnivores in relation not just to herbivores, but to different types of herbivore. A primate such as a chimpanzee, which is an eclectic feeder, taking nutritious foods such as insects, fruits and small mammals, has a much more complex environment and ecological world than a gorilla, which has been likened to a leaf-eater living in a salad bowl.

Size is another factor. Large animals require more food than smaller ones, and so they tend to need a larger area or home range over which to forage. These larger home ranges require a better sense of direction, a better capacity for navigation, and probably a better capacity for memory and prediction. A further factor might also be the length of time an animal lives. Animals that exist for only a short lifespan experience little in the way of environmental change, but those such as elephants, that may survive for 50 or more years, may experience many changes in the distribution of food, and hence must be more flexible, and responsive.

The inevitable conclusion would be that the most intelligent animals are those that are ecologically complicated. There is certainly considerable evidence to support this view. Among the primates, for example, if pairs of species that are closely related to each other, one of which has a complex foraging schedule, the other of which exploits ubiquitous food items, are compared, it is found that the former have the larger EQs. Gorillas, in their salad bowl, have EQs of 1.14; omnivorous chimpanzees have an EQ of 2.4. The 'lawn mowing' gelada baboon has a relatively small brain compared with the more generalized and frugivorous *Papio* baboon.[15]

The ecologists' answer is undoubtedly attractive, but has not been widely accepted. There is definitely an overall trend of larger brains being associated with more complex environments and foraging strategies, but there are many exceptions. Primate diets are not noticeably more complex and difficult to acquire than carnivore ones, and yet it is the primates that have the larger brains. Many species of rodent and bird manage to cope with very complex processes of searching for food and extracting it, with what amounts to relatively little neural tissue. Ecology is suggestive, but not conclusive. The second answer that has been proposed by Nicholas Humphrey and others such as Andy Whiten and Richard Byrne, is known as the social hypothesis.[16] The argument goes like this. Certainly finding food and avoiding predators is often a complex problem, but it is also relatively static. The appearance of fruits can be broadly predicted, and a lion soon learns that wildebeest have to drink water. Ecology is therefore relatively predictable, and does not require great intelligence, and in particular great flexibility. This can be contrasted with social interactions – that is, interactions between individuals of the same species.

Sociality is, like large brains, relatively rare in evolution. Most species live out a lone existence, meeting another member of the same species only once, fleetingly, for either mating or fighting. It would be wrong, though, to assume that sociality is uniquely human. The order primates is predominantly social, and there are other groups of mammals – for example the canids, or dog family – where sociality is extensively developed.[17] The impact that this has on evolution is fundamental.

First, though, it is necessary to clarify what is meant by sociality and being social. The most extreme interpretation of the term social is that it is the antithesis of what is biological. Things are either social or they are genetic and biological. This view, indeed, conflates the term biological to mean the same as genetic, a common fallacy within the social sciences.[18] This is a strongly anthropological and non-evolutionary perspective, which implies that sociality is a unique part of the human world. It ignores the fact that the capacity for social behaviour is predicated upon physical and biochemical, and indeed genetic characteristics, the widespread occurrence of social behaviour among other species, and the apparent trend towards marked sociality that can be found among the primates.

A second view is that the term social refers specifically to the presence of elaborate cognitive skills. Among humans the variation in social behaviour is exemplified by cultural diversity, which in turn implies a strong association between the social and the cognitive and symbolic systems that humans have developed. While the previous view took the term social away from that which is biological, within this second framework the concept of 'social' is removed from actual behaviour and focused on the cognitive capacities that generate behaviour. This view can be extended to include the idea that sociality is dependent upon the capacity for symbolic systems of thought and a sense of self-consciousness. While cognition is clearly an important element of any social behaviour, it is equally the case that in itself, it is not the totality. The social world is actually played out in the realm of behaviour.

This perspective on social cognition is strongly linked to the general notion of culture in anthropology; indeed, the terms culture and social are often used in an almost interchangeable manner. Culture, it could be argued, is concerned with the transmission of ideas and information through non-biological means, and as such it is transmitted largely through social channels. This view conflates the cultural capacities of humans with the tendency towards social life: to be social requires non-biological means of passing on information, and culture in turn requires individuals to live in social groups. There is thus a positive feedback between the two.

From the zoological end there are other notions of what is meant by the term social. For some biologists it simply means group living – that is, a social species is any species where individuals do not lead solitary lives, and all groups can be considered as social groups. Clearly this greatly extends the meaning of the term, and takes the problem of social evolution away from anything very special to include an enormous variety of biological problems, from clonal colonies of bacteria to shoals of fish. While there may be advantages in this view, the generality is too great to have much explanatory power; while many species may aggregate into groups of conspecifics, it is not necessarily the case that their interactions are in any meaningful sense of the word social. It is probably the case that the majority of social species live in groups, but it is not the case that all groups are social.

It is obvious that any attempt to model or explain social evolution is highly dependent upon the way in which the term social is used. In adopting those that are more specifically anthropological and which exclude biology, the power of any evolutionary approach is greatly reduced. Expanding the definition to all associations loses the focus on that which makes humans, and other primates, unique.

The solution adopted here is that formulated by Robert Hinde.[19]

Sociality is seen not as a top-down system, imposed by the characteristics of the group as a whole, but rather as an emergent property
derived from the interactions between individuals. The term 'social
groups' thus refers to groups where associations are maintained over
time and space, where the individuals are consistently interactive,
where individual recognition of others can be found, and where associations are patterned by familiarity and genetic relatedness. The
interactions themselves typically can be classified into a number of
simple categories. When these categories of interaction are patterned
over time, repeated between individuals, and both the context and
the content of the interaction are replicated, then social relationships
emerge from simple rules of interaction. These relationships can also
be patterned, maintain a stable context and content, and produce a
form of social structure which is unique to that cluster of individuals. Such an approach does not ignore the perceptual, communicative
or cognitive elements inherent in transactions; rather, in non-
linguistic species it provides a basis for definition and comparison. It
also stresses the distinction between the mechanisms by which
sociality is maintained (for example, cognition, behavioural cues,
etc.) and the actual behaviour of being social. Indeed Hinde emphasizes the dialectic between the individual, the relationship and the
emergent system. Again, as Lee[20] has argued, the nature of the relationships, which themselves influence interactions, and so on,
provides for a non-linear approach to understanding the complexity
of sociality, at least for species with 'maintained sociality'.

The complexity of social relationships provides the link to brain
evolution. Predicting the behaviour of another individual is hard,
especially if that individual's behaviour is itself predicated on what
another individual might be predicting. The flux of relationships pre-
sents a constant problem of updating behaviour and expectations in
line with experience and motivation. The calculus involved might
make the task of working out where to find the mongongo nuts pale
into insignificance.

Humphrey[21] has put this view most strongly, arguing that it is
sociality that drives the evolution of intelligence, and in consequence
brain size. The implication would be that the function of intellect, to
quote Humphrey, is to solve social problems. That it does other
things is a bonus. Byrne and Whiten[22] have taken this further by
focusing on one particular element of social interaction – that of
deceit. They have suggested that sociality is best thought of as a kind
of arms race between individuals trying on the one hand to predict
each other's behaviour, and on the other to avoid having their own
behaviour expected. Deceit, deviousness and surreptitiousness

would follow – the so-called Machiavellian ape. Dunbar[23] has pursued this theme quantitatively. He has plotted the number of primates in a social group against the ratio of the size of the neocortex (the 'thinking' part of the brain) to the rest of it. What this showed was that the larger the social group, the larger the brain. Where a monkey or ape had to remember its relationship with a large number of other individuals it required a large brain in order to do so. Or perhaps, as the direction of causality is not established by such a correlation, where an animal has only a small brain it is constrained in the size of group in which it can live.

This would seem to provide an answer, that the benefits of brains are to be found in social activities, and that the larger the social group, the greater the benefits, and therefore the greater the selective pressure on brain evolution. Other species do not have large brains because they do not live in social groups, and therefore they do not need them.

The Energetics of Intelligence

The conclusion that is drawing closer is that the benefits of intelligence for an animal are considerable, and while they are partly ecological, they are perhaps more convincingly social. But such a conclusion leads back to the original problem – if the benefits are so great, why are not all social animals as intelligent as humans? Why is there not a continual and increasingly rapid evolutionary race towards larger and larger brains. Why the evolutionary rarity?

This problem has arisen because, in terms of the model developed at the beginning of this chapter, so far only one side of the equation has been explored. We have looked at the benefits of large brains, but not the costs. Looking at the costs means returning to ecology and energetics.

Brains have a number of benefits, but they are also extremely costly. Metabolically neural tissue is extremely expensive.[24] The body can be thought of as a series of organs, necessary for life, but which cost energy for growth and maintenance. Some tissue is relatively 'cheap' – that is, it does not require great amounts of energy. Skin, bone, muscle, are cheap. Other tissue, such as the liver and kidneys is metabolically expensive.[25] Clearly the differences relate to the complexity of the functions they perform. Among the most metabolically costly tissues is the brain. The human brain is only 3 per cent of body weight, but it uses around 20 per cent of the energy required by an individual for metabolic maintenance.

The high energetic cost of brains helps resolve another problem –

the allometric relationship between brain size and body size discussed earlier. It was found that brains became larger at a lower rate than overall body size. In fact the rate was about three-quarters of body size increase. This rate is exactly the same as that for metabolic rate, which also scales allometrically with body size. In other words, brain size and body size are isometrically related, and brains appear to be energetically constrained.[26]

When considering brains and ecology the problem was put the wrong way round. It was not that a large brain was necessary for complex foraging, but that a high-quality diet, one that provided a stable and good source of energy and protein, was required to maintain a large brain. This explains the relationship between brain size and diet. A comparison of primate species shows this. Take any branch of the primates, and consider the species with the highest and lowest-quality diets respectively. Invariably, the species with the highest-quality diet has a higher EQ than the one with the poor diet – the chimpanzee versus the gorilla, siamangs versus gibbons, fruit-eating monkeys versus leaf-eating monkeys.[26]

When costs are taken into account, the rarity of the human evolutionary phenomenon is at last understandable. Certainly there may be benefits in large brains, both socially and ecologically, and there are selective pressures in favour of an increase in brain size, but these are usually outweighed by the costs. Most animals are better off putting their energy into muscle or large stomachs. There is in fact no conflict between the ecological and social elements in explaining the evolution of intelligence in the animal kingdom. The selective pressures favouring brain enlargement might be to do with the complexity of the environment, or social complexity, or a combination of the two, but the conditions that will allow the benefits to exceed the costs must ultimately be energetic. Such an argument is not only attractive because it fits the available evidence, but also because it achieves the aims set out at the beginning of this chapter – explaining not only why humans have evolved their unique characteristics, but also why other animals have not.

What is unique among humans as an evolutionary event is not the selective pressures leading to greater social complexity and more and more elaborate intelligence, but the occurrence somewhere in their evolutionary history of the ecological conditions that relaxed the constraints operating on other species and allowed the benefits of greater intelligence to greatly outweigh the costs. It is these ecological circumstances that are rare.

9

Why Did Humans Evolve?

The Final Problem

The conclusion reached at the end of the previous chapter represents the culmination of the Darwinian approach. Thinking about the costs and benefits and the balance between advantages and disadvantages in the way humans evolved, led to a recognition that humans could only evolve under specific conditions. How we became human was a product of the interaction between the populations of individuals that became our ancestors and the environments in which they lived and died.

This conclusion in fact leads directly back to the fundamental implications of the Darwinian revolution – Darwin's legacy of chapter 2. The first of these was that the problem of humanity was a scientific and technical one. Evolutionary theory is a specific theory that accounts for why particular organisms arise, survive and become extinct. The explanation it provides is that organisms are adapted to their environment, and therefore that certain conditions prompt certain results. This is the same as all scientific theories, which generally state that under certain conditions particular outcomes will occur. Under certain specific conditions of temperature and pressure water will boil; under other conditions it will freeze. Evolutionary theory is no different from these theories in physics, and it has been possible to show that under certain conditions large brains or bipedalism will be adaptive and will evolve, and under other (and indeed most) conditions, they will not. The basic question of palaeoanthropology, therefore, is nothing more than 'under what conditions will the human phenotype be adaptive?'

Additionally it should, of course, be added that while the general rule that evolutionary biology is like physics and chemistry in attempting to explain the conditions under which different outcomes may occur, there is a major practical difference, as we saw in Chapter 6. The particular outcome of any evolutionary situation will be dependent upon the interaction between the selective pressures and the environment in which the organism is living. But an additional and essential component is that the existing character of the species involved will have a major effect. The balance between costs and benefits will differ according to different phylogenetic or evolutionary histories. This does not mean that evolution is more than the interaction between environmental conditions and evolutionary players, but that the phrase 'other things being equal' seldom applies as each species has a unique history.

The second child of the Darwinian revolution was a belief that natural selection is the ultimate explanation for existence. What this means is that biological phenomena occur because they give their bearers a reproductive advantage. In the light of what has been learnt about human evolution, this means that it is necessary to set up an account of human history not in terms of who is related to whom, but in terms of how particular features gave hominids and humans advantages over contemporary competitive alternatives at particular times. Once again, the qualification that holds is that the calculus of benefits and costs is not an absolute one, but must be considered in the context of the specifics of the time and place in which each new feature arose.

The third consequence of the Darwinian revolution was the expectation and subsequent discovery of fossil hominids. These are the extinct species that fulfil several roles in the evolutionary framework. They are the missing links that show that humans are not entirely separate from the rest of the animal kingdom. They are the markers of the path by which an ape could become a human. They are also the evidence for the trial-and-error nature of evolution, as well as being the context in which later hominid and human characteristics had to establish their competitive credentials. Most important at this stage, though, is that they are the species, populations and individuals that we can at least partially fix in time and space, and hence use to test out the theories of how and why we became human.

The final problem will be to draw together the disparate strands that have ranged over such topics as biogeography, climatic change, ecology and the evolution of thermoregulatory systems, as well as the influence of historical factors on evolutionary change. The basis for such a synthesis lies in two particular elements. The first of these is that evolutionary change is prompted by the costs and benefits of particular strategies, mediated through the mechanism of differential reproductive success or natural selection. As such external environmental conditions do play a major role in evolution, and supply an ecological baseline from which all evolutionary explanations must develop. The second, which has become clearer as this book has progressed, is that evolution is not just about changes in anatomy and the hard biological tissues, but is based on the actual behaviour, the daily lives of the organisms themselves. Each creature plays an active as well as a passive role in evolution. For humans, therefore, to determine the ultimate causality of evolution it is necessary to examine the evolution of the sociality that underlies the expansion of the brain, in the context of the broader environmental conditions. Linking these factors together would constitute a Darwinian history of the human species, and would provide an explanation for the occurrence of the hominid lineage on this planet at a particular time and a particular place.

Social Evolution

Large brains, it was argued, are associated with being social, but the question of why animals should be social or what form that sociality should take has not been considered. If so much that is unique and special in humans is linked to sociality it is necessary to place this within an evolutionary framework.

The reasons underlying sociality are controversial, but as with large brains themselves there are both costs and benefits involved.[1] Being social means that there is more competition for food, but on the other hand there are more individuals who can search out food and perhaps co-operate in its acquisition. Social animals are likely to be more visible to predators, but equally there are more eyes to watch for predators. Social animals will compete more continuously for access to food and mates, but then again, they will often have the support of kin. The calculus of advantages over disadvantages will underlie the question of when animals will be social.

Sociality is not a unique human trait, but is integral to the whole of anthropoid evolution. The primates are the social order *par excellence.* Where other mammalian groups have specialized in large tusks or long necks, the primates have specialized in being social.[2] By living in groups, by developing sustained social relationships, they have survived and thrived over the aeons since their origin. Of the approximately 150 species of monkey and ape, only one does not live in some sort of social environment.[3] The one exception is the orang utan, which is solitary. There is evidence, however, that its solitary life is to some extent an illusion. While it lacks day-to-day contacts with other individuals, in fact it exists in a spatially widespread network of individuals which it knows and interacts with on occasion. In some ways the social demands of the orang utan are possibly more demanding than those of the more overtly gregarious and excitable species whose social relationships are updated from hour to hour.[4]

Sociality is really part of the primate core adaptation. It is, in the jargon of evolutionary systematics, a 'plesiomorphy' or primitively retained trait, rather than a unique feature of hominids and humans. It is the role of primatologists rather than anthropologists to explain the origins of sociality and society, for by the time hominids came along it was well and truly established.

If sociality itself is ancient within primate evolution, particular types of social organization and strategy are not, and it is the variation in sociality that is critical. It is often assumed within anthropological circles that social variation is infinite, and that the scope for human creativity in social relationships is massive. Zoologically this is not the case, for in practice there are only a limited number of ways of organizing a society, and some are far rarer

Primates are above all else highly social animals. Virtually all anthropoid primates live in complex social groups, and thus the human propensity of sociality is part of a more general primate trend.

than others. Recognizing that social options are finite is a useful start as it permits the narrowing down to clear finite limits both of what is socially possible and of the conditions under which various permutations can occur.[5]

The notion of 'finite social space' developed by Lee[6] is predicated upon a biological constraint that occurs in all mammals including humans. This constraint can be simply stated as the differential costs of reproduction to males and females; male gamete production is relatively inexpensive in terms of time and energy, and a male's reproductive potential is limited by his access to potential mates. Females, on the other hand, produce more energetically costly gametes, bearing the energy and time costs of internal gestation and subsequent lactation. A female's reproduction is thus largely limited by time and energy, through her condition and nutritional intake. In using this basic premise, the biology of reproduction is being placed at the core of sociality, a justifiable position given the central role played by reproduction in evolutionary mechanisms.

This model builds on Wrangham's original premise for primate sociality.[7] Females, constrained by the high costs of reproduction, must place the highest premium on ensuring access to sufficient energy for reproduction and parental care. The availability and distribution of resources across the landscape will therefore ultimately determine the optimal way in which a female should locate herself.

A female is constrained by energy and access to it. For males it is very different. Costs of reproduction to them are low, and what matters more than access to food is access to females. While females will distribute themselves around the food sources, males will distribute themselves around the females. The differences in the aggregations that arise constitute the variation, a finite variation, in social strategy and organization.

The options for social strategy can be specified with respect to the relationships between and within the sexes. A female may be solitary, she may associate with other females to whom she is related, or she may aggregate with other females to whom she is not related. Males have the same options in relation to other males: they can either be alone, with male kin or with male non-kin. When these strategies for within-sex relationships are meshed together it can be seen that the overall set of structures is relatively small. To these options can be added those of associations between the sexes. Males and females can associate in ways that are stable – forming associations that go beyond a single courtship and mating event – or the associations may be transitory. Degrees of stability may of course occur, including lifetime permanence.

These social structures have been deduced from a few simple principles, and are thus theoretical, but Lee[8] has shown how this account can describe the range of variation in mammals and the relative frequency of the various systems. Among a sample of 167 primates that have been studied sufficiently, 83 per cent are social rather than solitary, and all but one of the asocial species are prosimians rather than anthropoids. Among the social species, the most common form of social system is one where female kin form the basis for the social group (47 per cent). These are primate societies where females associate with other related females (female kin-bonded), and the males associate with them in circumstances either where there is a single male (the so-called harem) or where there may be a number of unrelated males. Monogamy is the next most common social structure (31 per cent). All other possibilities are comparatively rare or completely absent. There are for example at most nine social species (6 per cent) where male kin-bonding occurs, while polyandry (or female harems) is also known for about nine species. Totally absent for the most part are structures where both males and females associate with their kin, which means that a universal feature among primate species is that at least one sex will leave its natal group when it reaches maturity. Social groups are not, therefore, closed systems, but belong to larger population networks.

Furthermore, Rodspeth[9] and others have shown that this model also accounts for the pattern of variation in human social systems. Excluding completely celibate units (monasteries, nunneries and solitary individuals such as hermits), of the eighteen theoretically possible systems only six actually occur, the most common of which is conjugal families in larger communities of both male and female kin. Where kin of only one sex is involved it tends to be male.

The model of finite social space permits the range of social variation to be proscribed and described. It also opens up the possibility of considering social evolution, because by definition such evolution must consist of the transition from one state to another. It provides all the necessary components if it is possible to specify the ancestral state – that is, the influence of evolutionary history or phylogenetic heritage – and also the ecological conditions under which change may occur and new social options be adaptive.

Human evolution should be the specific test for this general model. So much of what makes humans special relates to their social behaviour that determining the 'cause' of human evolution must in some way require linking the ancestral social behaviour of the apes to the emerging environments in which the hominids found themselves.

A start can be made by looking at the phylogenetic context in

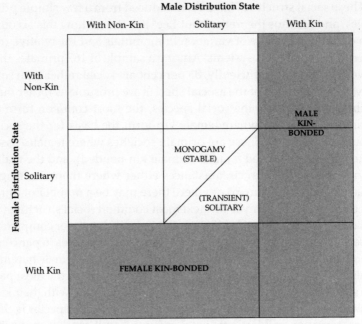

Male Distribution State

The finite social space model. The type of social structure any animal forms is based on how individuals organize themselves spatially relative to other individuals. According to the finite social space model the number of options are limited. Each sex may combine in certain ways with other members of its own sex, and with the opposite sex. Males may therefore associate with other males who are kin, or non-kin, or live alone. The same is true for females. Males and females may associate together either in transitory reproductive events, or more permanently represented here by a diagonal line in the central cell only.
The boxes show the finite possibilities of social structure that emerge, depending upon same sex and opposite sex associations.

which human sociality has evolved.[10] The living Old World anthropoids or Catarrhini are divided into two super-families, the Hominoidea and the Cercopithecoidea. The former comprise the apes and humans, the latter the monkeys (baboons, mangabeys, guenons and the leaf-eating colobines and langurs). While this evolutionary divide represents a number of anatomical differences, it also coincides with a marked difference in social strategy. In common with the primates as a whole, the monkeys, in the cases where kin-bonding occurs, are female kin-bonded. This means they are primarily characterized by male dispersal between groups and female residence. Where kin-based alliances occur, they are based on matrilines. In terms of the model described here, males are either solitary with respect to other males or co-resident with non-kin males.

In contrast to the typical female-kin residence of most Old World monkeys, among the Hominoidea stable female-kin residence is unknown. Both males and females can disperse – removing any potential degree of kin association – as in the monogamous gibbon and the solitary orang utan. Among gorillas the females disperse prior to reproduction, as do the majority of males, although some males remain resident with their fathers and these can ultimately inherit the harem. The common chimpanzee females generally disperse, while males remain co-resident with their male kin and form strong kin-based co-operative alliances. The situation among the pygmy chimpanzee or bonobo is interesting, in that females probably disperse and males remain co-resident, but the alliance structure seems to be concentrated on the male–female relationships rather than focused within the male kin units.

This social distinction is striking. The cercopithecoids are socially conservative – that is, they have the pattern found most frequently among social mammals – while the apes are both diverse and have a tendency towards the relatively rare condition of male kin-bonding. This poses the question of what sort of external conditions prompt either male or female kin-bonding, or in other words, what is the adaptive basis for different social strategies?

The Eternal Triangle: the Ecology of Social Life

Social groups will occur when the benefits exceed the costs, and many of these costs and benefits relate to the distribution and abundance of food. It is this that underlies what may be thought of as the ecology of social life. In trying to develop an idea of such an ecology, though, it is necessary to stop using the species as the unit of description, and to go down a level and examine the problem in terms of individuals, and in particular, in terms of the two sexes. What must be established is an ecology of being female and an ecology of being male.

Social groups will form when resources are clumped in large patches or where they are very uniformly and evenly spread across a landscape, allowing for a number of individuals to exploit the resources jointly. The important element is that the resources are extensive enough to be shared between a number of individuals without a significant reduction in individual intake. As shown earlier, it is the relationship between females in particular that is critical. Where resources occur in small dispersed patches, and the food in those patches is of a relatively high quality, then females will benefit from being solitary or asocial. The reason for this is that each individual patch would be too small to support more than one individual, and 'one patch, one primate' would be the best strategy. As the clumps become larger, especially if they are patchily distributed across the landscape and have high quality food, then joint exploitation becomes possible. Under these circumstances aggregations and social groups of related females should occur. In this case females have the benefit, through their relatives, both of the presence of other individuals to defend the resources, and of sharing them between individuals who share the same genes. In contrast, where the food is evenly distributed across the landscape, and is of a low quality, then the social aggregations are more likely to be of unrelated females.[11]

When high-quality food occurs in large patches, groups of related females may form, both to exploit and to defend the resource. Under such conditions, groups of related females thus suffer a smaller individual cost from the partitioning of resources amongst kin (reduced over the actual costs by the degree of relatedness), while gaining the considerable advantage of being able to control the resource against females who are not co-resident or related. Clearly, co-feeding need not lead to sociality, but in the presence of others who are in competition for that same resource, individuals who maintain sociality will gain an advantage. If the resources are widely enough spread, then the males may actively group females, who will tolerate co-feeding as there are few costs (because the food is widely available), and distinct advantages occur from staying with one or more males.

This leads on to the question of the ecology of males. Although food remains important, from an evolutionary point of view it is access to females that is more important. Where females seldom increase their reproductive success by increasing their access to males, for males the number of females they can impregnate is crucial. Females are the limiting resource of males. It follows, therefore, that where females are themselves highly dispersed (and solitary), then males too must be dispersed. This results either in a totally solitary or asocial condition, or else, where the females or a particular territory are defended, in monogamy. In extreme cases where both females and resources are rare this may develop into a polyandrous state of affairs. Where females are clumped around rich resources, then the males will compete for access, and a multi-male system with unrelated males is derived. If the females are highly clumped then a single male may be able to defend a harem. Where females are moderately dispersed, then such female defence becomes impossible and territorial defence may develop, where there is then the potential for that rarest of phenomena, male kin-bonding.

It can be inferred that the distribution and quality of resources influences the nature of social groups, and so in turn the pattern of evolution. It is necessary, however, to qualify somewhat the relatively hard ecological determinism that has just been outlined. Once social groups do form, then sociality will itself become one of the conditions influencing the costs and benefits of forming groups, and social evolution may therefore develop its own internal mechanism. It is likely to be the case that sociality may be one of those phenomena in evolution where once something has evolved, going back will be hard for selection to accomplish. Further social evolution would then be the product of social factors, but still predicated upon both the initial and the sustaining ecological conditions.

In considering the role of social factors in the evolution of larger brains we were able to establish two sets of relationships. The first was that there is a strong relationship between brains and social living – a positive relationship in which the more complex the sociality or the larger the group, the larger the brain could be expected to be. The second was that there is a relationship between brains and food quality, or with ecology more broadly. This is not the relationship which was first proposed, that brains were necessary to exploit certain types of resource, but that brains were metabolically expensive, and therefore their evolutionary existence was dependent upon a reliable source of high-quality foods. What this discussion has now established is the completion of the triangle of relationships – the link between sociality and ecology. This triangle

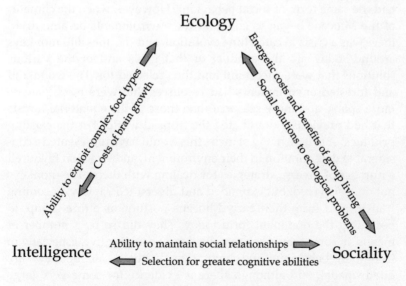

The triangle of relationships between sociality, intelligence and ecology.

allows the nature of complex evolutionary events to be unravelled; ecological conditions prompt responses in the form of social strategies, which in turn are dependent upon the ability of individuals to process information, an ability which in its turn is dependent upon various energetic constraints and therefore leads back to those self-same ecological conditions. Such a triangle shows why there cannot normally be an ever upward drive towards more complex social relationships and greater intelligence. It also tempts us into unravelling the sequence of events in the history of the one lineage where this did occur.

The Evolution of Human Social Behaviour

The key starting point for any history of human social behaviour should be the divergence of the apes and the monkeys over 25 million years ago.[12] Although this may seem rather remote, it represents a significant transition. The monkeys, as has been seen, have social systems in which female kin groups form the core of any social group where kinship matters. Although the females sometimes disperse and sometimes stay in their natal groups, the males invariably disperse. Where there are kinship systems, as there are in most baboons and macaques, they are what may be described as matrilines. Mothers, sisters and daughters are the permanent element. Males are merely transient members.

Early apes, and indeed most of the apes that have existed, probably had the same form of social behaviour. However, when the climates of the Miocene began to cool and the environments became drier, there was a crisis in catarrhine evolution, and the apes and monkeys around today are the product of that crisis and of the various solutions that were successful and thus selected for. The woodland and forested environments had resources that were more patchy, more sparse and more seasonal than those of the equatorial forests that had previously dominated the tropical world. For the existing primates, or at least many of them, this would have represented a considerable deterioration in their environment, and selection favoured a number of diverse strategies for dealing with them. The monkeys not only survived, but expanded and diversified rapidly, becoming transformed from their early Miocene position as a rare group, to becoming the dominant form today. They did so by a number of means. In some ways they remained very conservative, especially in locomotor behaviour and body size. Virtually all monkeys are quadrupedal, and although there is evidence for some very large baboons in the Pleistocene, for the most part they do not and did not exceed 30 kilograms. In other ways they developed a number of novelties, the most important of which was a tolerance for a wide variety of foods, the ability to process small hard seeds and fruits, and a capacity to detoxify the various defences that plants put up against herbivores. All this meant that they were able to expand their diet to include many of the more unpalatable and less nutritious of the resources in these new environments. Others developed stomachs and teeth that enabled them to live off large quantities of leaves.

In addition it is likely that the environmental crisis of the Miocene left its mark on the social behaviour of the catarrhines. The new resource structure would not have favoured particularly large groups, and the Miocene may well have been a time when kin-bonded forms of any sort became a rarity. Monogamous and solitary primates may have been more numerous. Only with full adaptation to the new environments in the Pliocene and Pleistocene, the period when the monkey species outnumbered ape species by two or three fold, did the large social groupings associated with the baboons in particular probably become the cercopithecoid norm.

The strategy of survival for the apes was very different. The diversity of apes decreased at the same time that the monkeys became so successful. Essentially those that did evolve and survive through the last 20 million years became larger. Again, in contrast to the monkeys, they elaborated a number of very new ways of moving around, becoming much less quadrupedal and much more suspensory in their behaviour. These developments in size and shape were offset by the fact that for the most part they remained conservative with regard to habitat and diet. Where the monkeys have adapted to the open savannas and even in the case of some baboons and the patas monkey, the desert, the apes have stayed largely within the forests or at the forest edges.

Despite some ecological conservatism, the social system seems to have been radically modified because body size has changed so markedly. The large body size did not favour the re-establishment of female kin-bonding. Instead, if the living apes are anything to go by, larger body size either put enormous strains on the very persistence of sociality, as was the case for the orang utan, or else it drove it in new directions. Where the monkeys seem to have established a basic and stable primate version of core mammalian sociality, the apes have stretched the whole nature of it to breaking point. This can be seen in the diversity of ape social systems, from the solitary orangs through the monogamous gibbons through the harem-holding gorillas to the friable and fissioning communities of the chimpanzees. The configuration of most interest to emerge out of this social manifestation of an ecological crisis is the male kin-bonding of the African apes.

Why should male kin-bonding emerge? Given the precedence that female strategy seems to have over male social strategy, the answer must lie in the conjunction between the larger size of the apes and the options available for females to form groups. The lack of female kin-bonded groups suggests either that the females are better off being solitary, or else that they can form groups with no costs and no benefits accruing to them. Size seems to be the key here. The larger females, depending upon small patches of food, widely dispersed, need to retain a flexibility in foraging that is incompatible with tightly bonded female groups. The fission and fusion within a larger community that occurs among chimpanzees reflects the costs of larger size – and larger size, it should be noted, is itself an adaptation that allows an animal to depend upon a more diverse range of food types.

The apparent indifference of females to the formation of kin-based groups opens up the options for males, and it is here that male kin-bonding can become established. Females may cluster in response to the behaviour of the males, and equally the males may benefit from the ability to exclude other males. Here alliances based on kinship can become advantageous, and as is found in chimpanzees the highly fissile internal community structure is strongly contrasted with the hostility which exists between males of the different communities. A further way in which male-based kinship systems may evolve is as a consequence of another outcome of increased body size. Large individuals tend to live a long time, and for males this means that their potential for increasing their reproductive success goes up drastically. Among monkeys, for example, a male will seldom maintain dominance for more than two or three years, and then he will start on the slippery downhill slope towards old age. For a gorilla, on the other hand, fifteen years of dominance is not unusual. This means that for a young male the chances of being able to usurp the place of another male and take over a harem is relatively small, and the risks of injury very high. Young males might well be better off remaining with their father and inheriting his harem. This in effect is a form of patriline, and the basis for male kin-bonding.

The emergence of male kin-bonding among the larger apes, and particularly in the African branch which includes the hominids, was probably a key event in evolution. It provides the ancestral condition from which the patterns of later hominids would be derived, and established the strategies that would be available for modification in the face of the further ecological and environmental shifts of the last

Hominoid Social Systems

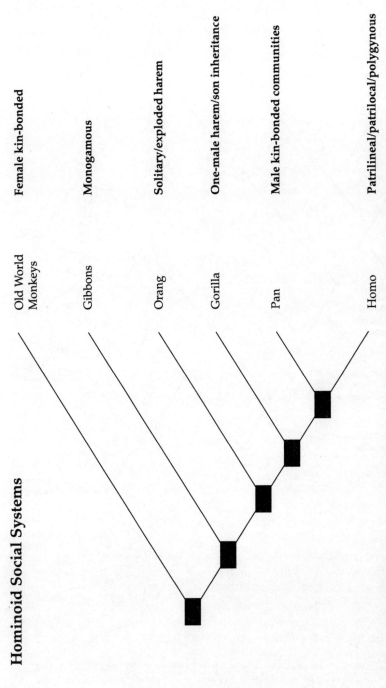

Female kin-bonded — Old World Monkeys

Monogamous — Gibbons

Solitary/exploded harem — Orang

One-male harem/son inheritance — Gorilla

Male kin-bonded communities — Pan

Patrilineal/patrilocal/polygynous — Homo

The evolutionary relationships of the apes and Old World monkeys, showing their social organization. What is striking is the variation in hominoid or ape social organization.

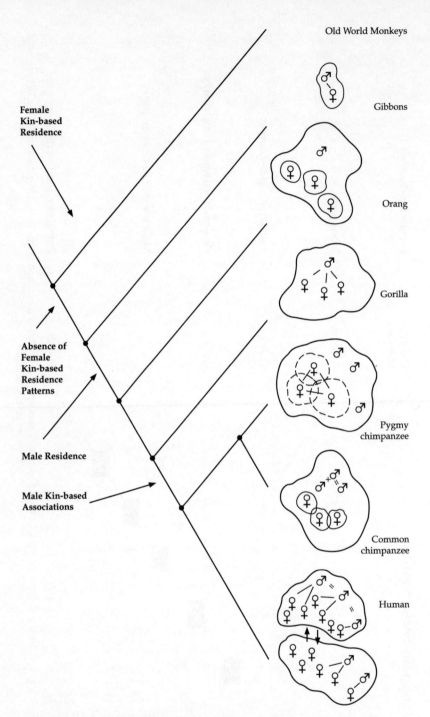

A model of the evolution of ape social behaviour and organization. The significant factor is the absence of female kin-bonding among the apes, and the development of male kin-bonding in the branch represented by chimpanzees and humans.

5 million years, the period of hominid evolution. It also illustrates the way in which behaviour, and social behaviour in particular, something normally beyond the reach of palaeobiology, is pivotal to understanding evolutionary events.

Hominids diverged from the other African apes in the later Miocene (9–6 million years ago), during a period of increasing climatic instability and aridity, and with further environmental shifts away from forests towards more mosaic woodland and grassland. The ancestral conditions for the African hominoids, and probably the common ancestor between chimpanzees and humans, can be surmised. In fully forested conditions there would have been a tendency towards closed, single-male groups as is found today among gorillas. Resources would have been abundant and distributed in large and uniform patches. Females would have been able to survive within small home ranges, and in turn single males would have been able to defend groups of females. In areas that were drier, and as forests became less continuous and were replaced by woodland and savanna, the resources would have become more dispersed and less uniformly distributed. Females would have had to forage more widely, and males would thus no longer have been able to employ a strategy of female defence. As is found among the common chimpanzee, larger communities would have occurred, with a breakdown of the classic harem system. With this, females would no longer have been attached to individual males, and related males would have had to co-exist to establish and maintain a territory. Thus, it is in the late Miocene that male residence, female dispersal, male kin-bonding and larger communities became established as the basic social organization of the clade leading to both chimpanzees and humans.

The last 8 million years have seen conditions becoming even drier and more seasonal. It is these conditions that are likely to have promoted the emergence of the most extreme ape of all, the hominid. In the context of the contrasting responses of monkeys and apes to these changes, the hominids may be thought of as the lineage that took the most radical departure from ancestral conditions, and in doing so managed to become the only ape to enjoy evolutionary success during the period when the monkeys were establishing themselves as the dominant group of Old World primates.

Bipedalism is the clearest evidence for the fundamental changes undergone by the hominids as they adapted to the eastern fringes of the African continent 5 million years or so ago. As was shown in chapter 7, these anatomical modifications represented the outcome of changes in behaviour, and especially changes in the way in which the first hominids foraged for food. However, these ecological changes are inextricably linked to changes in social behaviour.

The principal problem faced by the first hominids in the more open environments was that food was more dispersed and patchily distributed. It was the need to cover larger areas to find sufficient food that made bipedalism advantageous in hot, open environments. Predicting the effect of this on social strategy is not easy, but a number of suggestions might be made. Two contradictory ones are as follows. The first is that communities became smaller. It can be seen in chimpanzees that as they move into more arid environments their community size falls. In the wet and forested areas of Uganda chimpanzee communities are made up of between eight and one hundred individuals, whereas in the much drier Senegal they are around twenty-five.[13] In opposition to this, though, is the fact that among the real savanna primates, the baboons, groups are relatively large. This is probably a response to the increased threat of large predators.

Which strategy was pursued may have varied according to local conditions. Overall, though, it should be remembered that hominids arrived on the savanna with certain ancestral characteristics, including a tendency towards male kin-bonding, a community with great internal fissile capabilities and flexibility in day-to-day group structure, and hostility between communities. It may well be that the overall community size remained high to get the benefit of large numbers when threatened by conspecifics or large carnivores, but that these communities seldom operated as a whole and were both socially and spatially very disparate.

More important, perhaps, than overall group size, is its structure. There is little reason to believe that during the early stages of hominid evolution – the australopithecine grade – there was very great modification of the chimpanzee-style structure. The new environments offered few new options for female kin-bonded groups; plant resources would have been sparse and patchy, often unpredictable in both time and space, providing little that would have given them any benefits from forming kin-related groups. Furthermore, as the females would have been foraging over a large area, their distribution would have been too widespread to lead to the breakdown of multi-male groupings and direct female defence. Indeed, co-operative alliances of males may have been a major benefit that increased for ecological, territorial and anti-predatory reasons. Ecologically the male alliances may have been able to exploit the rich animal resources of the savanna environments; territorially they would have been able to patrol larger areas over which females and males would roam for food, and behaviourally they may well have been of at least some use in defence against the larger predators that were to be found in these more open environments. It can be argued that the potential for patrilineal kin-based social organization was in place in the last common ancestor of chimps and humans, and that its advantages were increased by the ecological and environmental shifts of the period from 10 to 5 million years ago – that is, the period of hominid origins. Variations in australopithecine social structure are likely to lie well within the range of possibility allowed by the African apes.

The Children of Human Evolution

If patrilineal affinal kinship relationships may be said to be African ape plesiomorphies stretching back to the Miocene, rather than unique human traits, where do relationships between males and females fit in? Close and long-term bonding between males and females, with prolonged parental and infant care, may not be universal among humans, but it is widespread, and furthermore it is totally absent from the behaviour of the great apes.

The model that has been developed in this chapter requires a consideration of this in terms of the factors that either increase or reduce reproductive success as a result of particular male strategies. If the chimpanzee condition is taken as similar to that of the common ancestor of hominids and other African apes, it is clear that neither males nor females would have gained any particular benefit from long-term close bonds. The social role of father did not exist, although this is not to say that males played no part in infant development. The implication might be that under new ecological conditions infant survival would have been more threatened, and greater care and effort may have been necessary to reduce the risks.

One such threat could have been a greater incidence of infanticide, and it may be the case that males were investing more to reduce the risk of infanticide by strange males. For the proposed model of a male kin-bonded social system this may seem unlikely, although the possibility of infanticide arising from hostile inter-group encounters cannot be excluded. Alternatively, paternal investment might have been increased if the infant survivorship became more and more dependent upon the quality of resources, and male protection or provisioning could have offset the risks. The key question becomes – under what conditions might this occur, and if so, when?

One possible factor may have been that hominid offspring became more costly. What this means is that their growth and development required greater levels of investment, in terms of either time or energy or both. The most probable cause of this would be the expansion of the brain among certain hominids.

In the previous chapter it was established that the brain is a metabolically expensive organ, requiring up to 20 per cent of energy expenditure. The larger the brain, the greater the costs. Larger brains therefore impose great costs, particularly on the mothers, for most brain growth occurs very early. In most mammals the brain achieves adult size while the animal is still very young, often before weaning has occurred. This means that the energetic costs of large brains are borne by the gestating and lactating mother. Such a cost can have a major effect on her strategies of survival and reproduction. She is tied even more closely to the quality and reliability of resources than ever before. The price for human evolutionary success falls most heavily on the mother.

A number of strategies are available to her. One is that she can spend longer on each infant, allowing the additional energy requirements to be spread out over a longer period of growth. Such a slowing down of the rate of maturation does occur in human evolution. Obviously this necessitates having fewer offspring, so that the survival of each one becomes more critical than ever. This can be seen in apes generally, where the interval between successive offspring (the inter-birth interval) is as much as five years, whereas for Old World monkeys it is seldom more than one or two years. It is here, perhaps, that the strategy of the male may change. To ensure the survival of these costly offspring a greater level of paternal effort may be advantageous, and with this will go the maintenance of bonds between male and female at least over the time when reproductive costs are greatest. Close, and ultimately emotionally strong relationships between males and females are the probable outcome of having large-brained offspring.

This of course begs a couple of questions. One is why the large brains? The other is, how could they be afforded? The answer to these questions leads back to the triangle of relationships – brains, sociality and ecology. Increasingly open environments occupied by the bipedal hominids represented the primary new conditions, and posed new problems for the way these highly social primates survived. Widely dispersed foraging and constantly shifting internal group structures such as those discussed above may have imposed the sorts of social selective pressures that Humphrey and others have argued underlie brain enlargement, and the costs of the large brain thus selected imposed new pressures on reproductive strategies that led to major changes in the way in which males and females related to each other. Large brains are a response to greater social complexity, whilst those large brains, with their high energetic costs, will reciprocally alter the nature of social relationships.

This neat circle is all very well, but it ignores one further dimension – where did the additional energy come from? It can be calculated that simply because of the size of the human brain, the metabolic costs imposed on the mother are around 9 per cent greater during the first eighteen months of life compared with a chimpanzee.[14] An additional 9 per cent is no small amount to be secured, particularly in environments that were often unpredictable. Additional energy might have come in a number of forms: greater efficiency in extracting the food, or reduced energy expenditure. Tools may have achieved this end, but for the early periods of human evolution at least these seem to have been sparse and relatively simple. A more attractive possibility lies in the consumption of meat.

Most primates are vegetarian in the sense that plant foods provide the bulk of their diet. Many species are also opportunistic consumers of some animal foods on a small scale. Only the baboon and chimpanzee have been noted for their active hunting of game. Modern humans are unusual among primates in raising the level of dependence upon meat.

There has been considerable controversy over the role of meat-eating in human evolution.[15] Most of this has focused on the practice of hunting itself – when was it possible, how important was it, what behaviours did it imply, and is it a fundamental part of the human psyche? In recent years discussion has tended to focus on the relative importance of hunting versus the scavenging of carcasses from other species' kills.[16] What has not been considered so fully is the implications of meat as a food, regardless of how it was acquired. Meat is a very high-quality food, having a high energy and protein content. It also comes in large packages, so that when meat is acquired there may be large quantities of it. There are disadvantages too, such as the fact that it is relatively sparse and the animals themselves are often very attached to their own muscle tissue. However, in the relatively plant-poor environments of the early hominids of Africa, meat may have been the critical resource – the one that provided that extra 10 per cent that enabled hominids to escape the constraints imposed on other animals in their response to the selective pressures imposed by developing social complexity.

Three points might sustain the crucial role of meat-eating in allowing the expansion of the human brain. The first is that in most of the African environments in which hominids have been found, meat is most available in the dry season, the period when plant foods are at their rarest, therefore solving the problem of the real crunch time in evolution – the season when resources are scarcest. The second is that the first signs of any enlargement in the size of the hominid brain occurs around 2 million years ago, with the first members of the genus *Homo*.[17] This coincides with the first evidence for tool-making and the first evidence for the exploitation of mammals as a food resource. And finally, Aiello and Wheeler[18] have noted that brains are not the only expensive tissue. Others such as the liver, the guts and the kidney are also metabolically costly. One way of coping with the additional costs of the large brain would be to reduce the size of another organ. Martin[19] noted some time ago that the human gut is among the smallest in the primates, and that generally carnivores have much smaller guts than herbivores; they need less length of intestine to digest their food. In adopting a greater use of meat hominids would have been simultaneously gaining access to a high-quality resource and reducing the energetic requirements of food processing.

These links between brains, social behaviour and ecology show that the unique outcome that is human evolution is certainly a rare event, occurring under extremely specific conditions. None the less, it is the outcome of the same basic principles that underlie all evolution. The evolution of humanity is the result of the specific interaction of a species with a complex social history coming into contact with a new and rewarding environment. Out of such an interplay of conditions can be written the Darwinian history of humanity.

Life History

This book started by looking at the humans before humanity as species, then as ecological beings, and finally as intelligent and social creatures. This would seem to take the problem of human history away from the generalities of biology and the certainties of genes, and place it instead in the framework of behaviour, something much more mutable and flexible. This might appear to relax the role of evolution and the significance of natural selection. The very success of the evolutionary explanation leads to its own overthrow.

Such may be the case, but it is worth ending this chapter by considering briefly how the social and behavioural strategies deduced here are in fact stamped on the whole biology of the species. Behaviour, thought, even cultural choices, do not occur in a vacuum, but are built on some very important biological predispositions.

Take the brain as the obvious example. All the capacity for learning, for creativity, for emotion, for rational thought, is entirely dependent upon the way in which the brain operates as an information processor. As we have seen, it is also just another biological organ, albeit a costly and complex one. Culture is therefore dependent upon the brain. The presence of the large human brain is in turn dependent upon certain mothering strategies, upon long periods of gestation, lactation and care. The nature of the females' reproductive biology has been selected for this capacity, and the very growth patterns of the human infant are tuned to these needs. Once again, behaviour and biology are deeply entwined.

The main point is that the whole time scale of an individual's life, from birth to death, is connected to biological processes, and those processes are the product of natural selection. Changes in what are known as life-history strategies – the length of time individuals spend in the various phases of life (gestation, infancy, adolescence, adulthood, and overall lifespan) underlie nearly all the evolutionary changes that have been discussed here. Male alliances, it was seen, are more likely to occur where individuals live a long time, and thus there is a link between longevity and social strategy. During the course of hominid evolution there has been an increase in the length of time an individual takes to reach maturity, and this is related to parenting strategies. Even the relative size of males and females – sexual dimorphism – shows the inexorable link between biology and behaviour. Among most mammalian species the males are much larger than the females when there is intense and risky competition amongst males for access to females. During the course of hominid evolution the level of sexual dimorphism can be seen to decrease, and this implies the changes in both male-male relationships and male-female relationships that have been discussed here. To take one final example, among modern humans women lose their reproductive capacity before they reach the end of their expected lifespan. The menopause prematurely terminates reproduction. This might seem a curious product of natural selection, but it has been suggested that the menopause is an adaptive switch imposed on human females by their costly offspring. Because the human infant takes so long to reach maturity the human female is better off not having those extra offspring, but perhaps helping her sons and daughters – what Hawkes[20] has termed the grandmothering strategy.

There are many regularities in life history when we look across the whole range of the animal kingdom. Long life goes together with large brains and large bodies, rapid growth is associated with early reproduction, and so on. In some ways humans fit neatly in with these expectations, and in others they extend life-history parameters and patterns into uncharted evolutionary territory. What is important, though, is that the links between the life history of a species and its behaviour document the far-reaching tentacles of natural selection.

10

Does Human Evolution Matter?

Nature's Place in Man

It is traditional in books on human evolution to end by considering the way in which our evolutionary past is important to understanding the world today. This may be relatively easy if the book is one such as Ardrey's *African Genesis*[1] for its theme made it easy to argue that having evolved from a killer ape into a killer human, it was important to understand the violent aspect of human nature. For many traditional anthropology books it would be impossible, for they have frequently argued that the principal pattern of human evolution is the gradual release of humans from the embrace of natural selection and the reduction in the effect of the environment on humans. If humans shape their evolution more than evolution shapes humans, then the role of the past might be of interest rather than significance.

In the preceding chapters I have taken something of an intermediate line. On the one hand I have emphasized the relentless nature of evolutionary processes such that they are seldom or never escaped, but on the other I have not attempted to identify the fundamental characteristic that makes us human. Indeed, there has been no evolutionary revolution, no sudden switch, and so the significance of it all is by no means clear. Furthermore, I have avoided any discussion of those great rubicons, culture and language, thus either relegating humans to being just another animal or simply leaving these to be the things that make the difference at the end. It remains perhaps to ask and answer one more question which will certainly have been prompted by some of the more tedious stretches of this book – does it matter that humans evolved?

One way of looking at the implications of what has been discussed in the earlier chapters is to go back not to Darwin, but to his contemporary, T. H. Huxley. Huxley wrote a book in 1863 called *Man's Place in Nature*.[2] This examined the evidence for humans having an evolutionary relationship with other creatures, and the extent to which there was fossil evidence for the transition from them. The book was largely zoological and anatomical, and included much of the evidence and lines of reasoning that have subsequently become the basics of the subject. *Man's Place in Nature* is undoubtedly a classic of evolutionary biology, as important as Darwin's *Descent of Man* published in 1871, and it set the context for many of the themes in twentieth century palaeoanthropology.

Huxley's basic question was 'where do humans fit in?' This is essentially a phylogenetic question, one about evolutionary relationships, and it is this question that has dominated the study of human evolution. It is the one that has been constantly revised with each new fossil discovery. Underlying it, and the title of Huxley's book, is the notion that the diversity of life is like a jigsaw, with a place for every species. Humans are an essential part of that jigsaw, without which it would be incomplete.

However, the implication of most of the preceding chapters has been that this is round the wrong way. Evolution in general and human evolution more specifically is much more localized, full of gaps and missing pieces, and more shapeless than Huxley conjectured. It is the processes and mechanisms, the links between environmental conditions and ecological and evolutionary outcomes, that are more central than where we fit in terms of evolutionary relationships. A Darwinian perspective poses the question of nature's place in man, rather than man's place in nature, for it is concerned with the way in which nature – or the environment and ecology, the conditions of evolution as we would now call it – have shaped humans. Humanity's place in the larger picture is simply an accidental outcome. By turning away from issues relating to the age of fossils and the role of particular species in human evolution, the focus has become the adaptive strategies of the hominids. This has meant that the heart of Darwinism – the mechanism of natural selection and the adaptations of populations – has become central, and evolutionary change, the usual centre stage, has been recognized as nothing more than an inevitable outcome of the changing interactions between organisms and their environments.

Culture Replaces Biology

One possible view of the significance of all that has been discussed here is that there is none. This might be the general view of what Tooby and Cosmides have called the SSSM – the Standard Social Science Model.[3] Social science in this century, as was argued in Chapter 1, has largely been built on the rejection of Darwinian theory and of an evolutionary perspective. While lip service may be paid to it having happened, the fact that it all happened so long ago has been sufficient reason to believe that its implications are minimal. A number of suggestions may be made as to why this is the case, ignoring for the moment the most obvious one that social scientists either dislike or do not understand evolutionary biology.

The first reason that has been suggested is that humans are not subject to evolutionary mechanisms because their behaviour is not subject to genetic control. Ernest Gellner has formulated this most elegantly with his theory of genetic underdetermination.[4] Humans are the way they are because their genes do not determine their behaviour, but rather, permit great variation and flexibility. In this way they have developed great powers of adaptability, able to survive and thrive in countless social and ecological circumstances.

Without genes, it is argued, there can be no selection, and therefore no evolution. The source of variation becomes cultural rather than biological, and therefore subject to a different set of rules. Humans would thus have made the transition from the world of biology to that of culture. As the power of genes has receded, Gellner has argued, culture has become essential as a means, and a better means at that, of reimposing constraints and rules. For a genetically under-determined species culture is a necessity for survival, but biology in turn becomes redundant.

This Standard Social Science Model amounts to a victory for the nurture side of the nature–nurture debate. Humans are the product of their cultural environment. There is certainly considerable evidence that is consistent with this view. The flexibility of human behaviour, the scale of cultural variation, and the extent to which individuals can adopt the practices of any culture would seem to be supportive of a largely environmentalist view of human nature. To this can be added the biological observation that most genetic variation is within populations, and therefore that humans are seldom genetically adapted to particular environments. Certainly it has not been possible to identify specific genes that control particular aspects of behaviour, and attempts that have been made are often little more than racial or sexual stereotypes.

If all this is true, then there is little in human evolution other than its curiosity value to those who like fossils. But is it true? Certainly there is a strong resurgence in genetics today, and behavioural genetics is a growing discipline.[5] This is seldom genetic determinism in a simple sense, for the genes are complex and it is clear that genes are mediated through any number of other factors. None the less there is a hereditary component to some features. Intelligence as measured by IQ, for example, shows levels of heritability that cluster around 0.7, which means that around 70 per cent of the variance is influenced by hereditary factors, while 30 per cent is due to what can broadly be called environmental ones.[6]

It is important to stress that this is by no means the same as saying that humans are the hapless product of their genes, for the environment modifies the influence of genes in a number of ways, and behaviour can never be allocated to nature or nurture; it is an inter-action process, with different elements having different functions. Genes influence the way in which the characteristics of an organism develop, be they physical or psychological, but the environment represents another source of influence, and one that affects the way in which the input from genes is made. In neither case can this influence be simplified to 'cause', for as we have seen in other contexts with evolution, causation is a mixture of factors and conditions.

Furthermore, it is by no means the case that genes are crucial to all aspects of the process of evolution. It is certainly a common mis-understanding that the terms biological, evolutionary and genetic are synonymous. In Dawkins's terminology,[7] genes are the replica-tors of evolution, the currency in which we measure whether change has taken place or not. Organisms, though, are the vehicles of replication, or in the terms of Hull,[8] the interactors with the environ-ment. Genes, after all, do not interact with anything, they simply code for phenotypes. This means that a gene's success is dependent upon the adaptive characteristics of the individual or vehicle. As Dunbar has said, 'Replication makes it possible for the organism's phenotypic activities to have the evolutionary consequences they have, but it is not what causes those consequences.'[9] Genes are necessary to evolution, but they are not the only thing that evolves, and the relationship between genetic and phenotypic evolution is not necessarily one-to-one. Elements of the phenotype may evolve relatively independently of any particular gene, although the whole genic package will benefit.

One implication of this is that genetic underdetermination is more an illusion than a reality from an evolutionary point of view, even if it usefully describes what we see in humans today. Rather than being unimportant, genes for behaviour occur at a very low level of specificity. There is no 'gene for aggressiveness' or 'gene for altruism'. What is far more likely is that there are genes that control for very general characteristics such as the ability to learn, or observe, or alter responses, and so on. This means that in the course of the evolution of human behaviour it is not specific behaviours that have been selected for, but the ability to respond appropriately to specific conditions. Certainly these may be built upon what Hinde has called biological predispositions,[10] but it means that evolution is central to understanding human behaviour without there being any simple one-to-one genetic component. Selection is still going on.

A further criticism of the Standard Social Science Model might be that culture is not a useful concept, and hides the problem rather than solves it. Culture is one of those concepts that is frequently used but seldom defined. Or perhaps more accurately, constantly redefined. At one extreme it simply means, in Lord Raglan's memorable phrase, 'everything that we do that the monkeys don't'. In this sense it is simply a tautology when used in relation to evolutionary or biological problems, for culture means those things that are not biological. At this level it is probably not very useful, and as genes do not work in isolation from other factors, neither is culture likely to.[11]

More specifically culture may mean one of two primary things. One would be the bundle of features which are possessed by humans, and which, if possessed by other species, occur only to a limited extent or are poorly developed. Such features would include the capacity to learn and teach, non-genetic transmission of information between and among generations, high levels of intra- and inter-population variability, the development of local traditions, the modification of the environment for technological purposes, and the use of symbolic elements of communication. The other would be that all these traits stem from the ability of humans to construct a mental world, and so culture would be the mental abilities that generate these observable phenomena.

The capacity for the Standard Social Science Model to account for humans rests ultimately on the adequacy of the culture concept. As an analytical and evolutionary concept this adequacy can be questioned. For one thing, it is a static, either/or concept. Creatures either have culture or they do not, hence for example the often tautologous discussions about whether chimpanzees or other living animals are cultural. Resolving this issue does not resolve anything about human evolution, because either chimpanzees are not, in which case the term simply maintains Lord Raglan's definition, or else they are, in which case some new definition of human uniqueness is required. The humans before humanity described here are the proof of the inadequacy of culture as an analytical tool, for it is clear that the extinct hominids lie exactly on the watershed of what is generally considered 'culture'. The complexity of the behaviours they were capable of cannot usefully be reduced to culture, and it has been possible to learn far more by avoiding a cultural approach altogether.

This does not preclude the possibility, however, that what has been bundled together as 'culture', especially the mental constructs that underlie the complexity of human behaviour, do not have a major impact on the way evolution may operate, as models developed to account for gene-culture evolution demonstrate. These models, however, reduce culture to a very much more limited and specific concept than is generally employed in the Standard Social Science Model.

An often stated contrast frequently used to underpin the rejection of evolution as a relevant theory is that humans have developed many behavioural aspects that have no function. Natural selection and evolution should produce better ways of eating or running away from predators, not better artists. Culture is therefore the things that are not really essential for survival, and so beyond the scope of evolution, or even acts counter to it. The hominid evolution that has been documented here has certainly been about the essentials of survival, but it has also demonstrated two other points that contradict this view.

The first is that natural selection is concerned with reproduction, not survival. Food, ecology, economics within a Darwinian framework are a means to an end, not an end in themselves. That end is strategies that ensure better reproductive opportunities, or better offspring survivorship. Anything that enhances this will be selected for, and it may well be that once the complex web of social and power relationships of the hominids was established, the sources of reproductive success were more than just simple foraging skills. The distinction between functional and non-functional elements has to be treated with great care in the context of evolutionary theory.

The second point is that the discussion of the relationship between brain size, sociality and maternal energetics in the preceding two chapters showed that it is not possible to separate the social from the ecological, the behavioural from the energetics on which they are based. Energy permeates all aspects of life, and therefore even the most functionally remote behaviour has the capacity to diffuse back into the more basic elements of life.

If culture and genetic underdetermination are inadequate reasons for rejecting evolution as a significant factor in understanding humans, then a second possible reason might be language itself. Language changes all the rules. A post-modernist extreme might simply be to say that none of the things discussed in this book can exist without language, therefore it is all just a language-filtered world view. Evolution does not matter because evolution is just a word. At a lesser extreme, language means that information is passed too rapidly to be influenced by or have an influence on evolution. Language has the effect of transferring the potential for adaptive and reproductive success from the individual to the group. At the very least this would mean that selection, which in classical evolutionary theory operates on the individual, will now operate at a higher group level, and that it is group costs and benefits that are now critical. This would undermine the various calculations that have been made throughout this book.

The evolution of language is often cited as the key step in evolution, allowing both the cumulative success of humans and the change in the pattern of evolution. Opinion differs as to when and how this occured. To some, it evolved gradually from the first appearance of the genus *Homo* around 2 million years ago. To others, it had a much more rapid origin, possibly as late as 40,000 years ago, associated with what is known as the symbolic explosion – the appearance of art, and of signs of the external use of symbols.[12] The runaway spread of modern humans, it has been argued, reflects the advantages that a language-using hominid has over a silent one.

But the role of language in human evolution must be treated with great caution. Language itself is of course excellent evidence for the relatively straightforward operation of biological evolution, since the capacity for language and speech is firmly rooted in biological properties – the structure of the brain and the anatomy of the throat, mouth and respiratory system. The benefits would therefore lie with the individual rather than with society as a whole. Furthermore, as Pinker has outlined recently, when viewed from an evolutionary point of view language is all the more remarkable.[13] There appears to be strong evidence for a universal structure to grammar, such that it is not just the physical capacity for speech that is the product of evolution, but the mental capacity too. Furthermore, language is built on an independent ability to think. Language is not necessarily part of thought processes, as images and dreams show. What Pinker refers to as 'mentalese' may be separate from language and more fundamental in evolution, with language being a significant addition.

Language, and the cognitive capacities that underlie it, are not reasons to ignore an evolutionary approach. Certainly language and human cognitive abilities have transformed the evolutionary history of the hominids. Mapping the evolution of the cognitive characteristics of the humans before humanity remains an important problem that needs to be tackled.

The Evolution of Human Cognition

Describing the nature of human cognition is entering a minefield of terminological and theoretical confusion. The way a human thinks can be considered at a number of levels, from the functioning of brain cells to the biochemistry of the brain, to its anatomical structure, to psychological and conscious components. These are not necessarily in conflict, but reflect the difference between the neurobiological processes underlying thought and the process of thought itself. It is the latter that is most easily framed in terms of the questions asked here – what are the costs and benefits of human cognitive capacity? The context for this is the evolutionary advantages of greater social complexity which we established in the previous chapter. The implication would be that the mind is adapted to charting courses through this complex social world and solving a range of environmental problems.

Humphrey has argued that the ability of the mind – any mind, not just the human one – has evolved from what may be considered to be primitive emotions.[14] An organism's awareness of its environment is derived from its senses – feeling pressure or temperature or light – and the development of increasingly conscious thought may be considered as arising as a means of making sense of these feelings. 'Thought' may therefore best be considered as a system of cognitive structures that organize responses to stimuli that arouse 'emotions'. There are important implications in this evolutionary model for understanding how the brain works, for it implies that there is a phylogenetic reason for the close links that occur between human emotions and rational thought. There is no basis for the view that there is a cold, calculating computer program of pure human thought independent of emotional states. Like everything else in evolution, even the power of the human mind has been built in a ramshackle way by adding new bits to existing components.

Two inferences about the evolution of human cognition can be made thus far. The first is that the way in which the brain operates is likely to be strongly linked to social processes, as these are the underlying selective pressures, and secondly that it will evolve in an additive way on existing capacities. The first of these in particular has been the basis for suggesting that a key stage in cognitive evolution is what is referred to as a theory of mind. This does not refer to a theory of how the mind works, but the idea that the mind works by having a theory that there is a mind. Whiten means by this that high levels of cognition involve using the process of thought to simulate activities, actions and consequences.[15] 'What if I hit the rock?' or 'What if I threaten X?' Thought is a means by which imagined consequences can be played out without expending the energy or taking the risk involved in the action itself. I might imagine hitting the rock and it breaking in half, with a nice sharp edge. Or I might imagine threatening X and realizing I would be in considerable danger because X is twice my size. The function of the mind is therefore to simulate actions. Because of the first of the principles already derived – that there will be a strong social element to the evolution of cognition – Whiten would see this primarily as mind reading. The most important simulation would involve trying to simulate what is going on in the mind of another individual. In imagining what would happen if I threatened X, my imagination, to be of any use, would have to take the form of trying to think what X would think of my threat. In effect this means getting into the mind of another individual, which can be both rational (a thought process) and emotional (a form of empathy) given, as was argued above, that there are no clear boundaries between different forms of cognition, but that one is built on the basis of the other.

It is this additive nature of the evolutionary process that may have been significant in the way human cognition evolved. What lies at the heart of the processes of thought is the ability to simulate actions mentally. While its origins may have lain in social events and problems, there is no reason why the power of the technique should have been confined to sociality. Its extension into the world of technology, ecology and many other realms of behaviour would have been an advantage.

One consequence of this model of cognitive evolution would be that it would lead to consciousness and self-consciousness. Any thought process which involves mentally manipulating the actions both of the thinker and of other individuals would logically entail an awareness of both self and others. It is clear from studies of other animals, especially the apes, that this level was achieved prior to the evolution of the hominids, and that the greater capacities that humans may have derived are from this anthropoid foundation.

The complexities of cognitive evolution summarized all too briefly in this way have formed the basis for an extraordinary range of work central to understanding humans. However, an issue tangential to most of this work is central here: can we map out the path of human cognitive evolution in the context of the conditions in which humans evolved? In this book it is not possible to look at the benefits without considering the costs, and considering the time scale involved in relation to the humans before humanity. We have learnt over the course of this book that it is not possible to assume that something like conscious thought is advantageous – the costs and the conditions are a necessary part of any explanation.

The evidence discussed in chapter 4 on the relative humanness of the various hominid taxa is central here. In looking at brain size, technology and ecological behaviour it was clear that the australo-pithecines were ape-like in everything other than gait, and that neanderthals had modern-sized brains but lacked modern behaviour. Indeed it is even the case that early *Homo sapiens* did not show 'modern' technology for over 60,000 years. Brains evolved late in hominid evolution, and there is a marked technological contrast between archaic and modern hominids. What does this tell us about the evolution of human cognitive systems?

First, that either the australopithecines were not under severe selective pressure for increased social complexity, or they were con-strained by energetic factors in the evolution of the neural and life-history parameters associated with later hominids. In all proba-bility their cognitive skills were within the range of those found in chimpanzees. Secondly the rate of technological change for the period from 2 million years ago to around 300,000 years ago (i.e. the disappearance of the Acheulean) is remarkably slow. Even the prepared core technologies of the Middle Stone Age and the Middle Palaeolithic show a remarkable stability over tens of thousands of years. This would again imply an absence of the characteristics of thought and language seen in modern humans.[16]

These two observations are consistent with the general trend of downgrading the capabilities of non-modern hominids and of em-phasizing the differences between them and *Homo sapiens*. This would suggest that the critical change occured late in hominid evo-lution, and was followed by a rapid explosion of cultural innovation. The problem with this interpretation lies in the energetic costs involved in larger brains, discussed in chapter 8. If non-modern hominids are doing so little by way of language and symbolic expres-sion, why should they have such large brains?

The solution to this apparent conflict that I would propose is the clear functional difference between thought and language.[17] Most attempts to understand the evolution of human cognition have made no such distinction, and have generally adopted the anthropological perspective that thought is internalized language. There are a number of reasons, though, for separating language, which relates to external communication, from what Pinker has referred to as 'mentalese', which is the internal thought processes.

If mentalese can evolve without an external manifestation of complex communication, then perhaps the gradual evolution of larger brains that occured during hominid evolution reflects this development. Its effect on the complexity of behaviour may have been profound, but without the rapid diffusion of ideas made possible by language, a relatively slow rate of behavioural change and a high level of stability occurred. The archaic hominids of the Pleistocene were probably adept mind readers, but they were also silent ones. Only over the last 300,000 years did brain evolution accelerate in such a way as to suggest that language itself might be evolving.

This suggestion, which is closely related to the evidence of the fossils themselves rather than to general models of cognition, has the advantage that it forces a consideration of the evolutionary factors involved. So far there has been a tendency to think in very general terms about cognition, but when placed into a Darwinian framework it is immediately obvious that there is no reason why selection for intelligence and mind reading is necessarily going to coincide with selective pressures to communicate the results of these thought processes. Selection for thought and selection for communication are two different processes. Indeed, there may be very good reasons why a lack of communication may be advantageous. What this implies is that during the course of hominid evolution there have been prolonged periods when the general benefits derived from greater powers of thought have been selected for, but such benefits did not accrue from language until the occurrence of some specific conditions in the last two or three hundred thousand years. Those conditions might well relate to the achievement of modern life-history parameters – the slow growth rates and greater levels of inter-dependence discussed at the end of the previous chapter. Alternatively it may result from the imposition of new conditions as the hominids accumulated greater population densities through the Pleistocene. Whichever is the case, it is a strong reminder that the benefits of language need to be demonstrated rather than assumed. Many hominids, including those with large brains, appear to have managed without its full modern manifestation.

The Nature of the Human Evolutionary Environment

The lesson of evolutionary biology over the last 50 years or so is that Darwinism does reach the parts that other theories do not. The extent to which Darwinism is about behaviour, the nature of social evolution, the flexible and context-specific aspects of so many evolutionary strategies, means that the pressure to include humans within the Darwinian framework is greater than ever before. It is this that is breaking up the Standard Social Science Model, and nowhere is this happening faster than in psychology.[18] The growth of evolutionary psychology has been dramatic, tackling from a Darwinian perspective such topics as the pattern of homicide, the mate preferences of males and females, and the nature of sexual choice. Even in anthropology the growth of behavioural ecology has produced fascinating results relating to the decline of polygamy, the nature of inheritance under different ecological conditions, and the role of inter-group violence in adaptive strategies. Those who have stayed with the book this far may not need persuading that this is hardly surprising, and it would be beyond the scope of this book to pursue it any further. Our concern has been the humans before humanity, rather than humanity itself. There is, however, a point of articulation.

Evolutionary psychology makes use of a concept known as the 'environment of evolutionary adaptedness' or EEA for short.[19] The EEA refers to the evolutionary environment that has shaped humans during the course of their evolution. While this need not necessarily be psychological alone, the implication is that there is a particular cognitive system that has evolved in response to the ecological and social conditions in which hominid ancestors found themselves. The nature of this environment is never very precise, but includes such generalizations as the hunter-gatherer way of life, or small-scale band society, and so on.

The EEA is clearly a potentially important concept, attempting to identify the evolutionary heritage that shapes the way we behave today. It has not, though, been without its critics even in sympathetic branches of anthropology, emphasizing as it does a unitary view of human evolution. Despite the link to the past, it plays down the importance of chronological patterns and minimizes the variation in human behaviour. The concept also flies in the face of one of the important conclusions of this book – that there is no single, invariant past, but an endlessly changing pattern of hominid populations responding to different conditions, and producing a diversity of adaptive strategies, only a few of which may be significant in the subsequent evolution of living human populations.

The fragmented nature of human evolution, made up of so many small events, makes it extremely hard to identify an EEA, or even a number of EEAs. And yet the spirit of the idea of the EEA is that the past does matter, that it shaped living populations and at many levels imposes constraints on the way people behave today. The problem is, can the details of human evolutionary history be resolved sufficiently to give a clear idea of the nature and depth of our evolutionary heritage?

The Darwinian Timetable for Human History

The significance of human evolution will vary from one person to another. An anatomist will have different concerns from a zoologist, and a zoologist will differ again from an anthropologist. An anatomist might be particularly concerned about the point at which there is conflict between pelvic outlet size and neonate headsize. A zoologist might be interested in the relationship between the level of behavioural difference and genetic distance in various species including humans, while a social anthropologist might want to know where the term culture becomes appropriate. Whatever the concern, though, the central point developed in this book is that while the past has an effect on the present, that effect cannot simply be left as a vague aura that hangs over us, but should be pinpointed to particular points in time and space. What this amounts to is a Darwinian timetable of human prehistory, or where and when did the benefits of particular human traits exceed the costs and provide a selective advantage. The list below represents a tentative attempt at such a timetable.

The anthropoid heritage (40 million years): This produced slow reproductive rates, leading to a tendency to have high levels of maternal effort and care. Sociality was firmly established as a core part of the primate adaptation, associated with a significant increase in brain size. The core would be strong bonds between mothers and offspring, with persistent relationships between females, especially kin-related ones, being built on these. Eclectic and opportunistic feeding with the ability to pick and choose resources became a common anthropoid characteristic. Cognitively the ability of individuals to recognize each other, to recognize kin, and to communicate simple intentions would have been present.

The hominoid heritage (25 million years): There was an absence of female kin-bonding as the core pattern of sociality, probably based on an inability to form large groups, and hence living in small 'nuclear' units.

The great ape heritage (15 million years): With great apes there was a loss of conservatism in life history parameters as a result of a tendency towards changes in body size as a solution to ecological problems. High levels of curiosity about the world, and greater dexterity and manipulative skills evolved, as well as great social flexibility in response to ecological conditions.

The African ape heritage (10 million years):   The heritage of the African apes is partly geographical, in that it was the vicissitudes of the changing African environment that provided the conditions for hominid evolution. As the centre of hominid distribution and the region where the hominids' closest relative lived, Africa may be considered the core demographic and selective environment. The other important element of the African ape heritage is the development of an essentially terrestrial or partly terrestrial way of life.

The 'Last Common Ancestor' heritage (7 million years):   The last common ancestor is taken to be the shared ancestor of humans and chimpanzees. The heritage here includes a number of key attributes, including an increased tendency to eat meat, the use of tools, including stone, to extract resources, and the active use of food as an element of sexual and reproductive behaviour and competition. Socially it is here that male kin-alliances become important, with the development of group defence and inter-group hostility between males. The other key element at this point is the fissile nature of the communities on a day-to-day basis, reflecting the way in which the larger communities are built on smaller nuclear ones. Cognitively there would have been an extensive understanding of social roles and relationships, considerable manipulation and manoeuvring, and the development of the arts of 'politics' – manipulating relationships and resources.

The australopithecine heritage (5 million years):   In all probability the primary difference in heritage between the australopithecines and the 'last common ancestor' is the adoption of bipedalism. This provides humans with a heritage involving the reorganization of the skeleton and the development of a number of physiological strategies for dealing with thermoregulatory problems.

The Homo *heritage (2.5 million years):*   The characteristics inherited from this stage would include the beginnings of brain expansion, the greater use of manufactured tools in foraging, and an increase in the use of animal food sources through hunting and scavenging.

The Homo erectus *grade heritage (1.8 million years):*   At this point there seem to be a number of very significant changes: the first real shift towards human life-history strategies occurs with a slowing down of growth rates, longer maturation, a much greater use of technology. The skills of observation and copying technology seem well established, although there is little evidence for innovation and variation. The capacity to maintain greater levels of gene flow over large areas develops, in association with the ability of the hominids to occupy diverse habitats and to colonize and disperse. It may well be from this point, with greater investment in offspring, that in addition to stronger male kin-alliances comes the beginnings of more exclusive and closer emotional bonds between males and females.

The 1000-gram brain heritage (300,000 years): The hominid brain reaches about 1000 grams at something under half a million years ago, and this seems to coincide with a number of changes. Most importantly the rate of evolution accelerates, especially in terms of cranial capacity. In association with this come a number of changes in technology, the beginnings of much greater regional and chronological variation in technological tradition, and greater apparent biological diversity within a grade. A number of theoretical reasons suggest that life-history strategies would need to be virtually the same as in modern humans for the larger brains to be sustained.[20] It is from this point on that something substantially different from non-human forms of communication should be assumed. Although the rate of behavioural change is not high, there is probably little doubt that these hominids had many of the traits that anthropologists would want to characterize as cultural forms of language may have been present.

The anatomically modern human (Homo sapiens) *heritage (140,000 years)*: The fully modern anatomy, although in a somewhat more robust form than is found in living populations, has an antiquity of over 100,000 years. It should, however, be remembered that this anatomy was not a single transition but evolved over a considerable period of time. It is likely that around the 150,000 year mark lies the last practical if not absolute common ancestral population of all living humans, and that this applies to the ancestor of modern languages as well.

The last 20,000 years: The period from 100,000 to 20,000 years ago saw the fully modern repertoire of behaviour come into place. However, it was only after 20,000 years ago that many of the things we associate with modern humans had their full impact. It was during this period that cultural diversification occurred, the consequences of which can still be seen. These human populations had an impact on resources and the environments that was significant, and a combination of environmental and demographic factors underlay the shift to food production which radically transformed the biological and cultural world. Socially and cognitively it is during this period that the large-scale social structures emerge and the socially extensive use of symbols becomes widely established.

It must be stressed that this timetable breaks the rules established through the rest of this book, that evolution must not be read as a process leading to humans. However, humans do exist, and this time scale does perhaps indicate how old various features are. What emerges most of all from this, perhaps, is that there is no complete package, but a mass of events building cumulatively on each other. What is unique in humanity is entirely dependent upon the level of comparison. For the humans before humanity this means that each of them will have different combinations of these traits, and as such will extend the range of variation of the biological world beyond anything we see today in the surviving lineages.

A striking omission from this timetable is much discussion of the time scale of the evolutionary differences between males and females. One reason for this is that the timetable only began with the anthropoids, and the ecological and evolutionary basis for the

50 Myr	Anthropoid heritage: compulsive sociality
10–5 Myr	African ape heritage: male kin-bonding
2–1.6 Myr	*H. erectus* heritage: expensive offspring life history strategy
300 Kyr	Archaic *sapiens* heritage: modern human life history strategy
150 Kyr	Modern human heritage: large group size
10 Kyr	Agriculture and the demographic revolution

A timetable for the evolution of human social behaviour.

different strategies of male and female, as was argued in chapter 9, are fundamental to sexual reproduction. What will therefore be fixed in any evolutionary heritage will be the sensitivity of the female sex to fluctuations in resources, and among males will be flexibility in strategies of access to females and the subsequent involvement with any offspring. Certain things may turn out to be strongly correlated with these biological predispositions, such as males being risk-prone and females risk-averse. The evolutionary heritage of sexual differences beyond these things, though, is likely to be the ability of males and females to deploy strategies of reproduction and offspring survival that are appropriate to specific ecological and social conditions. These are not fixed strategies, such as monogamy, polygamy, promiscuity and so on, but are better thought of as strategies for calculating the costs and benefits of greater or lesser degrees of sexual exclusivity, permanence of attachment between sexes, degree of parental involvement and so on. Biology and ethnography both indicate that males and females will sometimes be drawn to similar strategies, but the calculus of reproductive costs will mean that under many conditions they will not.

This timetable goes against a number of orthodoxies and current interpretations. One of the most popular is that there were significant events 40,000 years ago which led to what has been referred to as the 'human revolution'.[21] This was the point at which it is often thought that the 'symbolic explosion' occurred, when art flourished and human creativity and cultural complexity became established. The timetable I have outlined hardly mentions 40,000 years ago, for the simple reason that on a global scale, nothing much happened then. Forty thousand years ago (or more reasonably, between 45,000 and 30,000 years ago) there was a change in the population of Europe, the disappearance of the European archaic populations (the neanderthals) and the appearance of a new and more complex technology. Part of that is revealed by the presence of some beads, and one or two other items showing a high level of creativity. However, this must be put into a broader context; in many parts of the world there is no such change throughout the Pleistocene, whilst in Africa these features may be a good 30,000 years older. Even in Europe there is no sudden explosion, and the often cited cave art fluorescence in south-west France actually occured around 13,000 years ago – or, to put it in biological terms, over one thousand generations later. There is also a marked disjunction between 40,000 years and the appearance of anatomically modern humans (*Homo sapiens*), which is at least 100,000 years older.

My own view would be that if there was a period when something

212 DOES HUMAN EVOLUTION MATTER?

like a human revolution occurred it was much later, between about 15,000 and 5,000 years ago when there was a massive ecological and demographic change throughout virtually the whole world. This means that the evolution of humans did not stop 100,000 years ago with our anatomical form, nor 40,000 or 50,000 years ago with the Upper Palaeolithic of Europe, but was still happening when the world was undergoing the transition to agriculture. This takes us a long way from Darwin and Leakey's views of the great antiquity of the human lineage, stretching back 20 million years or more. Instead this is the short view with a vengeance. More accurately, of course, it is the expected conclusion from a process that is continuous and cumulative. It is also the predicted outcome of the principles of neo-Darwinism with which we started – the potential for evolutionary change will exist as long as there is differential reproduction between individuals and populations.

Fearsome Assymetry: Ancient Trivial Pursuits and Grand Consequences in Human Evolution

The consequences of the presence of humans on this planet are staggering. Since the end of the Pleistocene 10,000 years ago many environments and landscapes have been transformed by human activity. Deserts such as the Sahara have spread and continue to do so at terrifying rates. Forests have been totally destroyed, and the remaining ones are disappearing at a rate measurable in days. River systems have been reconstructed to suit human needs, and massive lakes created behind dams that have flooded thousands of square kilometres. Over the last 400 years the size of the human population has increased from a few hundred million to nearly 7 billion, and it is likely to increase further. There is growing evidence that human activities in the form of agriculture and industrial production are changing the climate. This is unintentional, but it may not be too long before there are active attempts to alter the atmosphere. For the last 10,000 years at least humans have altered the size and shape and behaviour of plants and animals through selective breeding, and they have in effect created new species. Genetic engineering and biotechnology will accelerate this process to the point where there will be the potential to change permanently every species. At the same time species are becoming extinct at an unprecedented rate. Every year it is estimated that thousands disappear, and the level of biodiversity will be reduced, possibly forever.

While humans have been having an impact on the world that is probably unprecedented, they have also amassed a history that goes beyond anything in nature. Cities, states and empires have risen and fallen, in the process of which they have amassed and controlled people in their millions. Weapons of fearsome destruction, as well as just the cumulative effect of endless wars and feuds between and within states, have left millions dead: 15 million people died in Russia between 1914 and 1922, and then another 20 million in the years of the Second World War, and these figures can be matched across the globe. In addition there has been a growth of science and technology such that diseases have been eradicated and the world reduced to the scale of a global village. And finally, humans have shown a capacity for creativity, selfless behaviour and heroism that makes the poverty, carnage and environmental destruction seem like a bad dream.

The phenomenon of man is quite a phenomenon whatever criteria are used and whatever scale is considered. Whether in terms of triumphs or disasters, there is nothing modest about Project Humanity. It is not surprising, therefore, that the existence of humans has inspired equally phenomenal explanations. It is only natural that big events should have causes commensurate with their effects. The dinosaurs are much the same; their extinction must have been cataclysmic and so deserves to be marked by the most massive asteroid collision in the Earth's history. It would simply not be right for them to dwindle into oblivion. Equally humans require something special, something to match their effects – a means to justify their ends. Theologians, philosophers, scientists and journalists have not been slow to produce such a means, and explanations for the existence of humans can take any number of forms. Many involve going beyond the Earth to existence somewhere else. This may be God or extra-terrestrial creatures, but the effect is to mark out the Earth for special creation. Humans are too improbable for them to have just occurred. Other explanations, particularly those in the post-Darwinian world, may involve events taking a remarkably long time, on the grounds that the highly improbable requires the bulk of geological time. More frequently, though, evolutionary explanations emphasize ways in which humans or their ancestors have departed radically and completely from other evolutionary trends, or else have escaped the normal constraints of biology. Killer apes, aquatic apes, symbol-using apes and lop-sided apes have all burst through the barriers of evolution – with one bound our hero was free.

What all these have in common is the imposition of the improbable, the freakish and the unusual onto the course of evolution. Sometimes this may occur in the name of random processes, a lucky mutation that changed the history of the world. At other times it is highly deterministic, as if the whole course of evolution needs humans to make sense of it all. The result, though, is always the same, the transformation of the ordinary into the extraordinary.

The oddity of all this is that stated baldly it is clearly illogical. It has to be at least admitted that there is no inevitable reason why large-scale and impressive phenomena should have large-scale causes. Only hindsight leads the eye and the mind to this conclusion, and hindsight is one privilege that evolution cannot have. Evolution must be read backwards, but it occurs forwards through time. In unravelling human evolution as I have tried to do here, it has been important to give up the benefit of hindsight. This has meant looking at humans through the lens of the fossilized and extinct hominids – the humans before humanity – rather than from the perspective of the twentieth century. This has, I hope, been revealing. There remains much that we do not know, and there is probably much that we never will. However, the larger patterns rather than the often obscure names can answer some of the larger questions. The nature of humans, and an understanding of why they evolved, has to be sought in terms of the past, of the hominids and the world they inhabited, and it is important that the massive and bewildering impact of their existence should not mislead us into confusing causes with consequences. As we have seen here, human evolution is no blinding flash and no special creation. Man did not make himself, nor woman herself. Both are the product of countless events in the daily lives of the hominids. There is no magic ingredient in human evolution, and no substitute for knowing the details of what happened – where and when and why. Small, insignificant earthquakes in Africa, or particular demographic trends in Europe, are responsible for what happened in evolution. We should not let the uniqueness of our species dupe us into believing that we are the product of special forces. Cosmologists studying the origins of the universe need to think in terms of a big bang. Evolutionary biologists are better off with a bout of hiccups. If we had been privileged enough to observe the origins of our species and our lineage, we would have been struck by one thing – nothing very much happened.

Notes

Chapter 1: A Question of Evolution

1 Ruse (1979 1986).
2 Maynard Smith (1989).
3 Zetterberg (1983).
4 Spencer, H. (1851).
5 Bowler (1989, 1990).
6 Kuper (1983); Harris (1968).
7 Bowler (1989).
8 Harris (1968).
9 Pinker (1994).
10 Ghiselin (1969).
11 Mayr and Provine (1979); Mayr (1982, 1991).
12 Gould (1980); Eldredge and Cracraft (1980); Gould and Eldredge (1977).
13 Eldredge and Grene (1992).
14 Dover (1986).
15 Gould (1989, 1991).
16 Fisher (1930).
17 Mayr (1963); Huxley (1942).
18 Wilson (1976); Hinde (1970).
19 Kimura (1983).
20 Barkow et al. (1992); Buss (1994); Whiten (1993).
21 Betzig et al. (1988).
22 Williams (1966).
23 Morris (1968).
24 Van Daniken (1968); Morgan (1982); Bokun (1979).

Chapter 2: Why Darwinism?

1 Hugh-Jones (1979).
2 Landau (1991).
3 Van Daniken (1968).
4 Morgan (1982).
5 Bokun (1979).
6 Medawar (1967).
7 Bowler (1989).
8 Dawkins (1986).
9 Bowler (1989); Ghiselin (1969).
10 Popper (1974).

Chapter 3: What are Human Beings?

1 Mayr (1991).
2 Wallace (1870, 1889).
3 Darwin (1871).
4 Darwin (1872).
5 Fisher (1930); Ridley (1993).
6 Goodall (1970, 1986).
7 Kortlandt (1986).
8 McGrew (1992).
9 Napier (1971); Le Gros Clark (1959).
10 Harcourt et al. (1981).
11 McGrew (1992).
12 Dart (1949).
13 Ardrey (1967).
14 Lee, R. B. and Devore, I. (1968).
15 Goodall (1986); Boesch, C. and Boesch, H. (1989).
16 Lieberman (1991); Bickerton (1990).
17 Oakley (1959).
18 Goodall (1970).
19 Goodall (1963); Harding (1973).
20 Gardner, R. A. and Gardner, B. T. (1969).
21 Foley (1995b).

Chapter 4: When did we Become Human?

1 Reader (1988); White et al. (1994).
2 Darwin (1871).
3 Bowler (1989).
4 Rudwick (1971).
5 Lewin (1987).
6 Lewin (1987).

7 Huxley (1863).
8 Dart (1925).
9 Spencer, F. (1990).
10 Washburn (1983); Spencer, F. (1985).
11 Stanley (1978).
12 Avise (1994).
13 Sarich and Wilson (1967).
14 Hoelzer and Melnick (1994).
15 Sibley and Alquist (1984); Williams and Goodman (1989); Horai et al. (1992).
16 Avise (1994).
17 Pilbeam (1967).
18 Avise (1994); Li and Grauer (1991).
19 Darwin (1871).
20 Lewin (1987); Reader (1988).
21 White et al. (1994); Wolde Gabriel et al. (1994); Wood (1994).
22 Hill and Ward (1988).
23 Johanson and Edey (1981).
24 Leakey and Hay (1979).
25 Aiello and Dean (1990).
26 Schick and Toth (1993).
27 Foley (1987b).
28 Foley (1995b).

Chapter 5: Was Human Evolution Progressive?

1 Johanson and White (1978).
2 McHenry (1992).
3 Martin, pers. comm.
4 Senut and Tardieu (1985).
5 White et al. (1993).
6 White et al. (1994); Leakey, M. et al. (1995).
7 White et al. (1983); Tobias (1980)
8 Grine (1989).
9 Walker et al. (1986).
10 Leakey et al. (1964); Tobias (1991).
11 Wood (1991).
12 Swisher et al. (1994).
13 Clarke (1985).
14 Klein (1989).
15 Walker and Leakey (1993); Rightmire (1990).
16 Swisher et al. (1994); Gabunia and Vekua (1995).
17 Bilsborough (1991).
18 Wood (1984); Andrews (1984).
19 Tattersall (1986).
20 Howell (1991).

21 Foley (1989a).
22 Foley (1991).
23 Kingdon (1984).
24 Gould (1989).
25 Skelton and McHenry (1992).
26 Grine (1989).
27 Wood (1991).
28 Wood (1991).
29 Groves and Mazak (1975); Groves (1989).
30 Stringer and Gamble (1993).

Chapter 6: Why Africa?

1 Leakey, L. S. B. (1961).
2 Weidenreich (1943).
3 von Koeningswald and Weidenreich (1939).
4 Diamond (1991).
5 Darwin (1871).
6 Foley (1987a, 1989a).
7 Behrensmeyer and Hill (1980).
8 Brain (1981).
9 Foley (1993).
10 Foley (1984, 1987).
11 Vrba (1985); Maglio and Cooke (1978); Turner and Wood (1993).
12 Jacob (1982).
13 Aiello (1981); Foley (1987a); Fleagle (1988).
14 Eisenberg (1981).
15 Myers and Giller (1988).
16 MacArthur and Wilson (1967); Brown (1981); Wright (1983).
17 Gamble (1994).
18 Foley (1987a).
19 Lodge (1993); Crosby (1986); Groves and Burdon (1983).
20 Mayr (1963).
21 Cann et al. (1987).
22 Mellars and Stringer (1989).
23 Lahr and Foley (1994).
24 Stringer and Andrews (1988).
25 Brauer (1992).
26 Frayer et al. (1993).
27 Tattersall (1986).
28 Foley and Lahr (1992); Lahr and Foley (1994).
29 Stoneking (1993); Melnick and Hoelzer (1993).
30 Cann et al. (1987).
31 Templeton (1993).
32 Cavalli-Sforza et al. (1988); Mountain et al. (1993).

Chapter 7: Is Human Evolution Adaptive

1   Maynard Smith (1974).
2   Lewis (1960).
3   Bateson (1988).
4   Darwin (1871).
5   Lovejoy (1981).
6   Foley (1987a).
7   Foley (1987, 1984).
8   McGrew et al. (1981).
9   Rodman and McHenry (1980).
10 Rodman (1984); Goodall (1986).
11 Morris (1968).
12 Hardy (1960).
13 Morgan (1982).
14 Roede et al. (1991).
15 Wheeler (1985, 1991a, 1991b).
16 Wheeler (1985, 1991a, 1991b).
17 Dunbar (1992).
18 Dunbar (1992).
19 Foley (1992).
20 Wheeler (1991b).

Chapter 8: Why are Humans Such an Evolutionary Rarity?

1   Jerison (1973).
2   Dawkins (1986).
3   Wolpoff (1971).
4   Kortlandt (1972).
5   Krebs and Davies (1991).
6   MacFadden (1992).
7   Krebs and Davies (1991).
8   Cheyney and Seyfarth (1990); Byrne and Whiten (1986); Whiten (1991).
9   D'Arcy Thomson (1942); J. S. Huxley (1932).
10 Jerison (1973); Martin (1983); Eisenberg (1981).
11 McHenry (1992).
12 Aiello and Dean (1990); Martin (1983, 1989).
13 Martin (1981).
14 Milton (1988); Clutton-Brock and Harvey (1980); Gibson (1986).
15 Foley and Lee (1991).
16 Humphrey (1976); Byrne and Whiten (1986).
17 Lee, P. C. (1994).
18 Ingold (1990).
19 Hinde (1976, 1983).
20 Lee, P. C. (1994).
21 Humphrey (1976).
22 Byrne and Whiten (1986).

23 Dunbar (1992a).
24 Martin (1981).
25 Wheeler and Aiello (1995).
26 Martin (1981).
27 Foley and Lee (1991).

Chapter 9: How do we Explain the Evolution of Humans?

1 Lee (1994).
2 Jolly (1972).
3 Smuts et al. (1985); Dunbar (1988b).
4 Rodman and Mitani (1985).
5 Foley and Lee (1989).
6 Lee (1994).
7 Wrangham (1980).
8 Lee (1994).
9 Rodspeth et al. (1991).
10 Foley (1989b); Wrangham (1987); Giglieri (1987); Foley and Lee (1989, 1995).
11 Wrangham (1980).
12 Foley (1989b); Foley and Lee (1989, 1995).
13 Foley (1993).
14 Foley and Lee (1991).
15 Dart (1949); Ardrey (1967); Lee and DeVore (1968); Dahlberg (1981); Foley (1987a).
16 Binford (1984); Bunn and Kroll (1986).
17 Foley (1987a, 1992).
18 Wheeler and Aiello (in press).
19 Martin (1989).
20 Hawkes (1989).

Chapter 10: Does Human Evolution Matter?

1 Ardrey (1967).
2 T. H. Huxley (1863).
3 Tooby and Cosmides (1992).
4 Gellner (1989).
5 Plomin et al. (1990).
6 Bouchard et al. (1988); Mascie-Taylor (1993).
7 Dawkins (1981).
8 Hull (1988).
9 Dunbar (1988), p. 181.
10 Hinde (1976).
11 Boyd and Richerson (1985); Durham (1991).
12 Falk (1980); Lieberman (1986); Bickerton (1990); Davidson and Noble (1991)

13 Pinker (1994).
14 Humphrey (1992).
15 Whiten (1991).
16 Foley (1995).
17 Foley (1995).
18 Buss (1994).
19 Symons (1979).
20 Foley and Lee (1991).
21 Mellars and Stringer (1989)

References

Aiello, L. C. 1981. Locomotion in the Miocene Hominoidea. In C. B. Stringer (ed.), *Aspects of Human Evolution*, London: Taylor and Francis, pp. 63–98.

Aiello, L. C. and Dean, M. C. 1990. *An Introduction to Human Evolutionary Anatomy*. London: Academic Press.

Andrews, P. J. 1984. An alternative interpretation of the characters used to define *Homo erectus*. *Courier Forschungsinstitut Seckenberg*: 167–75.

Ardrey, R. 1967. *The Territorial Imperative*. London: Collins.

Avise, J. C. 1994. *Molecular Markers, Natural History, and Evolution*. Chapman and Hall.

Barkow, L., Cosmides, L. and Tooby, J. 1992. *The Adapted Mind*. Oxford: Oxford University Press.

Bateson, P. P. G. 1988. The active role of behaviour in evolution. In M. W. Howard and S. W. Fox (eds), *Evolutionary Processes and Metaphors*, New York: John Wiley.

Behrensmeyer, A. and Hill, A. 1980. *Fossils in the Making*. Chicago, University of Chicago Press.

Betzig, L., Borgerhoff-Mulder, M. and Turke, P. 1988. *Human Reproductive Strategies*. Cambridge: Cambridge University Press.

Bickerton, D. 1990. *Language and Species*. Chicago, University of Chicago Press.

Bilsborough, A. 1991. *Human Evolution*. Glasgow: Blackie.

Binford, L. R. 1984. *Faunal Remains from Klasies River Mouth*. New York: Academic Press.

Boesch, C. and Boesch, H. 1989. Hunting behaviour of wild chimpanzees in Tai National Park. *American Journal of Physical Anthropology*, 78: 547–73.

Bokun, B. 1979. *Man: The Fallen Ape*. London: Sphere Books.

Bouchard, T. J., Lykken, D. T., McGue, M., Segal, N. L. and Tellegan, A. 1988.

Sources of human psychological differences: the Minnesota study of twins reared apart. *Science*, 250: 223–8.

Bowler, P. J. 1989. *Evolution: The History of an Idea*. Berkeley: University of California Press.

Bowler, P. J. 1990. *Charles Darwin, The Man and his Influence*. Oxford: Basil Blackwell.

Boyd, R. and Richerson, P. 1985. *Culture and the Evolutionary Process*. Chicago: University of Chicago Press.

Brain, C. K. 1981. *The Hunters or the Hunted: An Introduction to African Cave Taphonomy*. Chicago: University of Chicago Press.

Brauer, G. 1992. L'hypothèse Africaine de l'origine des hommes modernes. In J. -J. Hublin and A. M. Tillier, *Aux origines d'Homo sapiens*, pp. 181–215, Paris: Presses Université de France.

Brown, J. H. 1981. Two decades of homage to Santa Rosalia: towards a general theory of diversity. *American Zoologist*, 21: 877–8.

Bunn, H. T. and Kroll, E. 1986. Systematic butchery by Plio-Pleistocene hominids at Olduvai Gorge, Tanzania. *Current Anthropology*, 27: 431–52.

Buss, D. M. 1994. *The Evolution of Desire*. New York: Basic Books.

Byrne, R. and Whiten, A. (eds) 1986. *Machiavellian Intelligence*. Oxford: Clarendon Press.

Cann, R. L., Stoneking, M. and Wilson, A. C. 1987. Mitochondrial DNA and human evolution. *Nature*, 325: 31–6.

Cavalli-Sforza, L. L., Piazza, A., Menozzi, P. and Mountain, J. 1988. Reconstruction of human evolution: bringing together genetic, archaeological and linguistic data. *Proceedings of the National Academy of Sciences*, 85: 6002–6.

Cheyney, D. L. and Seyfarth, R. M. 1990. *How Monkeys see the World*. Chicago: University of Chicago Press.

Clarke, R. J. 1985. *Australopithecus* and early *Homo* in southern Africa. In E. Delson (ed.), *Ancestors: The Hard Evidence*, New York: Alan Liss, pp. 171–7.

Clutton-Brock, T. H. and Harvey, P. H. 1980. Primates, brains and ecology. *Journal of Zoology*, 190: 309–23.

Crosby, A. W. 1986. *Ecological Imperialism: The Biological Expansion of Europe 900–1900*. Cambridge: Cambridge University Press.

Dahlberg, F. (ed.) 1981. *Woman the Gatherer*. New Haven: Yale University Press.

Dart, R. A. 1925. *Australopithecus africanus*: the man ape of South Africa. *Nature*, 115: 195–9.

Dart, R. A. 1949. The predatory implemental technique of the *Australopithecus*. *American Journal of Physical Anthropology*, 7: 1–38.

Darwin. C. R. 1871. *The Descent of Man and Selection in Relation to Sex*. London: John Murray.

Darwin, C. R. 1872. *The Expression of the Emotions in Man and Animals*. London: John Murray.

Davidson, I. and Noble, W. 1991. The evolutionary emergence of modern human behaviour: language and its archaeology. *Man*, 26: 223–54.

Dawkins, R. 1981. *The Extended Phenotype*. San Francisco: Freeman.

Dawkins, R. 1986. *The Blind Watchmaker*. Harlow: Longman.

Diamond, J. 1991. *The Rise and Fall of the Third Chimpanzee*. London: Radius.

Dover, G. A. 1986. Molecular drive in multigene families. *Trends in Genetics*, 2: 159–65.

Dunbar, R. I. M. 1988a. The evolutionary implications of social behaviour. In H. C. Plotkin (ed.), *The Role of Behaviour in Evolution*, Cambridge, MA: MIT Press, pp. 165–88.

Dunbar, R. I. M. 1988b. *Primate Social Systems*. London: Croom Helm.

Dunbar, R. I. M. 1992a. Neocortex size as a constraint on group size in primates. *Journal of Human Evolution*, 22: 469–93.

Dunbar, R. I. M. 1992b. Time: a hidden constraint on the behavioural ecology of baboons. *Behavioural Ecology and Sociobiology*, 31: 35–49.

Durham, W. 1991. *Coevolution*. Palo Alto: Stanford University Press.

Eisenberg, J. 1981. *The Mammalian Radiations: An Analysis of Trends in Evolution, Adaptation and Behaviour*. London: Athlone Press.

Eldredge, N. and Cracraft, J. 1980. *Phylogenetic Patterns and Evolutionary Processes*. Columbia.

Eldredge, N. and Green, M. 1992. *Interactions: The Biological Context of Social Systems*. New York: Columbia University Press.

Falk, 1980. Hominid brain evolution: the approach from palaeoneurology. *Yearbook of Physical Anthropology*, 23: 93–107.

Fisher, R. A. 1930. *The Genetical Theory of Natural Selection*. Oxford: Oxford University Press.

Fleagle, J. 1988. *Primate Evolution and Adaptation*. Academic Press.

Foley, R. A. (ed.) 1984. *Hominid Evolution and Community Ecology: Prehistoric Human Adaptation in Biological Perspective*. London and New York: Academic Press.

Foley, R. A. 1987a. *Another Unique Species: Patterns in Human Evolutionary Ecology*. Harlow: Longman.

Foley, R. A. 1987b. Hominid species and stone tool assemblages: how are they related? *Antiquity*, 61: 380–92.

Foley, R. A. 1989a. The ecology of speciation: comparative perspectives on the origins of modern humans. In P. A. Mellars and C. B. Stringer (eds), *The Human Revolution: Behavioural and Biological Perspectives on the Origins of Modern Humans*, Edinburgh: Edinburgh University Press, pp. 298–320.

Foley, R. A. 1989b. The evolution of hominid social behaviour. In V. Standen and R. A. Foley (eds), *Comparative Socioecology*, Oxford: Blackwell Scientific Publications, pp. 473–94.

Foley, R. A. 1991. How many hominid species should there be? *Journal of Human Evolution*, 20: 413–27.

Foley, R. A. 1992. Evolutionary ecology of fossil hominids. In E. A. Smith and B. Winterhalder (eds), *Evolutionary Ecology and Human Behavior*, Chicago: Aldine de Gruyter, pp. 131–64

Foley, R. A. 1993. The influence of seasonality on hominid evolution. In S. J.

Ulijaszek and S. Strickland (eds), *Seasonality and Human Ecology*, Cambridge: Cambridge University Press, pp. 17–37.

Foley, R. A. 1995a. Language and thought in evolutionary perspective. In I. Hodder et al. (eds), *Interpreting Archaeology*, London: Routledge, pp. 76–80.

Foley, R. A. 1995b. The causes and consequences of human evolution. *Journal of the Royal Anthropological Institute*, 30: 1–20.

Foley, R. A. in press. Measuring the cognition of fossil hominids. In P. Mellars and K. Gibson (eds), *Modelling the Early Human Mind*, MacDonald Institute Monograph, Cambridge.

Foley, R. A. and Lahr, M. M. 1992. Beyond out of Africa: reassessing the origins of *Homo sapiens*. *Journal of Human Evolution*, 22: 523–9.

Foley, R. A. and Lee, P. C. 1989. Finite social space, evolutionary pathways and reconstructing hominid behaviour. *Science*, 243: 901–6.

Foley, R. A. and Lee, P. C. 1991. Ecology and energetics of encephalization in hominid evolution. *Philosophical Transactions of the Royal Society*, London Series B, 334: 223–32.

Foley, R. A. and Lee, P. C. in press. Finite social space and the evolution of human social behaviour. In S. Shennan and J. Steele (eds), *Cognition and Social Evolution*, London: Routledge.

Frayer, D. W., Wolpoff, M. H., Thorne, A. G., Smith, F. H. and Pope, G. G. 1993. Theories of modern human origins: the palaeontological test. *American Anthropologist*, 95: 14–50.

Gabunia, L. and Vekua, A. 1995. A Plio-Pleistocene hominid from Dmanisi, East Georgia, Caucasus. *Nature*, 373: 509–13.

Gamble, C. 1994. *Timewalkers: The Prehistory of Global Colonisation*. Stroud: Alan Sutton.

Gardner, R. A. and Gardner, B. T. 1969. Teaching sign language to a chimpanzee. *Science*, 165: 664–72.

Gellner, E. 1989. Culture, constraint and community: semantic and coercive compensations for the under-determination of *Homo sapiens sapiens*. In P. A. Mellars and C. B. Stringer (eds), *The Human Revolution: Behavioural and Biological Perspectives on the Origins of Modern Humans*, Edinburgh: Edinburgh University Press, pp. 514–28.

Ghiselin, M. T. 1969. *The Triumph of the Darwinian Method*. Berkeley: University of California Press.

Gibson, K. R. 1986. Cognition and brain size and the extraction of embedded food resources. In J. Else and P. C. Lee (eds), *Primate Ontogeny, Cognition and Social Behaviour*, Cambridge: Cambridge University Press, pp. 93–105.

Giglieri, M. P. 1987. Sociobiology of the great apes and the hominid ancestor. *Journal of Human Evolution*, 16: 319–57.

Goodall, J. 1963. Feeding behaviour among wild chimpanzees: a preliminary report. *Symposia of the Zoological Society of London*, 10: 39–47.

Goodall, J. 1970. Tool use in primates and other vertebrates. In D. S. Lehrman, R. A. Hinde and E. Shaw (eds), *Advances in the Study of Behaviour*, New York: Academic Press.

Goodall, J. 1986. *The Chimpanzees of Gombe*. Cambridge, MA: Belknap Press.

Gould, S. J. 1980. Is a new and general theory of evolution emerging? *Palaeobiology,* 6: 119–30.

Gould, S. J. 1989. *Wonderful Life.* New York: W. W. Norton.

Gould, S. J. 1991. *Bully for Brontosaurus.* New York: W. W. Norton.

Gould, S. J. and Eldredge, N. 1977. Punctuated equilibria: the tempo and mode of evolution reconsidered. *Palaeobiology,* 3: 115–51.

Grine, F. E. (ed.) 1989. *The Evolutionary History of the 'Robust' Australopithecines.* Chicago: Aldine de Gruyter.

Groves, C. P. 1989. *A Theory of Human and Primate Evolution.* Oxford: Clarendon Press.

Groves, C. P. and Mazak, V. 1975. An approach to the taxonomy of the Hominidae: Gracile Villafrancian hominids of Africa. *Casopis pro Mineralogii Geologii,* 20: 225–47.

Groves, R. H. and Burdon, J. J. 1983. *Ecology of Biological Invasions.* Cambridge: Cambridge University Press.

Harcourt, A. H., Harvey, P. H., Larson, S. G. and Short, R. V. 1981. Testis weight, body weight, and breeding system in primates. *Nature,* 293: 55–7.

Harding, R. S. O. 1973. Predation by a troop of olive baboons (*Papio anubis*). *American Journal of Physical Anthropology,* 38: 587–92.

Hardy, A. C. 1960. Was man more aquatic in the past? *New Scientist,* 7: 642–5.

Harris, M. 1968. *The Rise of Anthropological Theory.* New York: Columbia University Press.

Hawkes, K., O'Connell, J. F. and Blurton Jones, N. G. 1989. Hardworking Hadza grandmothers. In V. Standen and R. A. Foley (eds), *Comparative Socioecology,* Oxford: Blackwell Scientific Publications, pp. 341–66.

Hill, A. and Ward, S. 1988. The origin of the Hominidae. *Yearbook of Physical Anthropology,* 31: 49–83.

Hinde, R. A. 1970. *Animal Behaviour.* New York: McGraw Hill.

Hinde, R. A. 1976. Interactions, relationships and social structure. *Man,* 11: 1–17.

Hinde, R. A. 1983. *Primate Social Relationships: An Integrated Approach.* Oxford: Blackwell.

Hoelzer, G. A. and Melnick, D. J. 1994. Patterns of speciation and limits to phylogenetic resolution. *Trends in Evolution and Ecology,* 9: 104–7.

Horai, S., Satta, Y., Hayasaka, K., Kondo, R., Inoue, T., Ishida, T., Hayashi, S. and Takahata, N. 1992. Man's place in Hominoidea revealed by MtDNA genealogy. *Journal of Molecular Evolution,* 35: 32–43.

Howell, F. C. 1978. The Hominidae. In V. J. Maglio and H. B. S. Cooke (eds), *Evolution of African Mammals,* Cambridge, MA : Harvard University Press, pp. 154–248.

Howell, F. C. 1991. The integration of archeology with paleontology. Paper presented at the Spring Symposium, Field Museum, Chicago.

Hugh-Jones, S. P. 1979. *The Palm Tree and the Pleades.* Cambridge: Cambridge University Press.

Hull, D. L. 1988. Interactors versus vehicles. In H. C. Plotkin (ed.), *The Role of Behaviour in Evolution,* Cambridge, MA: MIT Press, pp. 19–50.

Humphrey, N. K. 1976. The social function of intellect. In P. P. G. Bateson and R. A. Hinde (eds), *Growing Points in Ethology*, Cambridge: Cambridge University Press, pp. 303–17.

Humphrey, N. K. 1992. *A History of the Mind*. New York: Simon and Schuster.

Huxley, J. S. 1932. *Problems of Relative Growth*. London: Methuen.

Huxley, J. S. 1942. *Evolution, The Modern Synthesis*. London: Allen and Unwin.

Huxley, T. H. 1863. *Man's Place in Nature*. London: Williams and Norgate.

Ingold, T. 1990. An anthropologist looks at biology. *Man*, 25: 208–29.

Jacob, F. 1982. *The Possible and the Actual*. Seattle: University of Washington Press.

Jerison, H. J. 1973. *Evolution of the Brain and Intelligence*. New York: Academic Press.

Johanson, D. and Edey, M. A. 1981. *Lucy*. New York: Simon and Schuster.

Johanson, D. C. and White, T. 1978. A systematic assessment of African hominids. *Science*, 203: 321–30.

Jolly, A. 1972. *The Evolution of Primate Social Behaviour*. London: Macmillan.

Kimura, M. 1983. *The Neutral Theory of Molecular Evolution*. Cambridge: Cambridge University Press.

Kingdon, J. 1984. *East African Mammals: An Atlas of Evolution in Africa*. Chicago: University of Chicago Press.

Klein, R. G. 1989. Extinction of the robust australopithecines. In F. E. Grine (ed.), *The Evolutionary History of the 'Robust' Australopithecines*, Chicago: Aldine de Gruyter.

Kortlandt, A. 1972. *New Perspectives on Ape and Human Evolution*. Amsterdam: Stichting voor psychobiologie.

Kortlandt, A. 1986. The use of stone tools by wild-living chimpanzees and earliest hominids. *Journal of Human Evolution*, 15: 77–132.

Krebs, J. and Davies, N. (eds) 1991. *Behavioural Ecology: An Evolutionary Approach* (3rd edition). Oxford: Blackwell Scientific Publications.

Kuper, A. 1983. *Anthropology and Anthropologists: The Modern British School*, London: Routledge & Keagan Paul.

Lahr, M. M. and Foley, R. A. 1994. Multiple dispersals and the origins of modern humans. *Evolutionary Anthropology*, 3(2): 48–60.

Landau, M. 1991. *Narratives of Human Evolution*. New Haven: Yale University Press.

Le Gros Clark, W. 1959. *The Antecedents of Man*. Edinburgh: Edinburgh University Press.

Leakey, L. S. B. 1961. *The Progress and Evolution of Man in Africa*. Oxford: Oxford University Press.

Leakey, L. S. B., Tobias, P. V. and Napier, J. R. 1964. A new species of the genus *Homo* from Olduvai Gorge. *Nature*, 202: 7–9.

Leakey, M. D. and Hay, R. L. 1979. Pliocene footprints in the Laetoli beds at Laetoli, northern Tanzania. *Nature*, 278: 317–23.

Leakey, M. G., Feibel, C. S., McDougall, I. and Walker, A. 1995. New four million year old hominid species from Kanapoi and Allia Bay, Kenya. *Nature*, 376: 565–71.

228   REFERENCES

Lee, P. C. 1994. Social structure and evolution. In P. Slater and T. Halliday (eds), *Behaviour and Evolution*, Cambridge: Cambridge University Press, pp. 266–303.

Lee, R. B. and DeVore, I. 1968. *Man the Hunter*. Chicago: University of Chicago Press.

Lewin, R. 1987. *Bones of Contention*. New York: Simon and Schuster.

Lewis, R. 1960. *The Evolution Man*. Harmondsworth: Penguin.

Li, W. -H. and Li, D. 1991. *Fundamentals of Molecular Evolution*. Sunderland, MA: Sinnauer.

Lieberman, P. 1986. *The Biology and Evolution of Language*. Cambridge, MA: Harvard University Press.

Lieberman, P. 1991. *Uniquely Human*. Cambridge, MA: Harvard University Press.

Lodge, D. M. 1993. Biological invasions: lessons for ecology. *Trends in Evolution and Ecology*, 8: 133–7.

Lovejoy, C. O. 1981. The origin of man. *Science*, 211: 341–50.

MacArthur, R. H. and Wilson, E. O. 1967. *The Equilibrium Theory of Island Biogeography*. Princeton: Princeton University Press.

MacFadden, B. J. 1992. *Fossil Horses: Systematics, Paleobiology, and Evolution of the Family Equidae*. Cambridge: Cambridge University Press.

Maglio, V. and Cooke, H. B. S. (eds) 1978. *The Evolution of African Mammals*. Cambridge, MA: Harvard University Press.

Martin, R. D. 1981. Relative brain size in terrestrial vertebrates. *Nature*, 293: 57–60.

Martin, R. D. 1983. *Human Brain Evolution in an Ecological Context*. 52nd James Arthur Lecture on the Evolution of the Brain, American Museum of Natural History.

Martin, R. D. 1985. Primates, a definition. In B. A. Wood, L. Martin, and P. J. Andrews (eds), *Major Topics in Primate and Human Evolution*, Cambridge: Cambridge University Press.

Martin, R. D. 1989. *Primate Origins and Evolution: A Phylogenetic Reconstruction*. London: Chapman and Hall.

Mascie-Taylor, C. G. N. 1993. Galton and the use of twin studies. In M. Keynes (ed.), *Sir Francis Galton FRS: The Legacy of his Ideas,* London: The Galton Institute, pp. 119–43.

Maynard Smith, J. 1974. The theory of games and the evolution of animal conflict. *Journal of Theoretical Biology*, 47: 209–21.

Maynard Smith, J. 1989. *Did Darwin Get it Right?* London: Penguin.

Mayr, E. 1963. *Animal Species and Evolution*. Cambridge, MA: Harvard University Press.

Mayr, E. 1982. *The Growth of Biological Thought: Diversity, Evolution and Inheritance*. Cambridge, MA: Harvard University Press.

Mayr, E. 1991. *One Long Argument*. Cambridge, MA: Harvard University Press.

Mayr, E. and Provine, W. B. (eds) 1979. *The Evolutionary Synthesis*. Cambridge, MA: Harvard University Press.

REFERENCES 229

McGrew, W. C. 1992. *Chimpanzee Material Culture*. Cambridge: Cambridge University Press.

McGrew, W. C., Baldwin, P. J. and Tutin, C. E. G. 1981. Chimpanzees in a hot dry, open habitat: Mount Assirik, Senegal, West Africa. *Journal of Human Evolution*, 10: 227–44.

McHenry, H. 1992. How big were early hominids? *Evolutionary Anthropology*, 1: 15–20.

Medawar, P. 1967. *The Art of the Soluble*. London: Penguin.

Mellars, P. A. and Stringer, C. B. (eds) 1989. *The Human Revolution: Behavioural and Biological Perspectives on the Origins of Modern Humans*. Edinburgh: Edinburgh University Press.

Melnick, D. J. and Hoelzer, G. A. 1993. What is MtDNA good for in the study of primate evolution? *Evolutionary Anthropology*, 2(1): 2–11.

Milton, K. 1988. Foraging behaviour and the evolution of primate intelligence. In R. Byrne, and A. Whiten (eds) 1986, *Machiavellian Intelligence*, Oxford: Clarendon, pp. 285–305.

Morgan, E. 1982. *The Aquatic Ape*. London: Souvenir Press.

Morris, D. 1968. *The Naked Ape*. London: Jonathan Cape.

Mountain, J., Lin, A. A., Bowcock, M. and Cavalli-Sforza, L. L. 1993. Evolution of modern humans: evidence for nuclear DNA polymorphisms. In M. J. Aitken, C. B. Stringer, P. A. Mellars (eds), *The Origin of Modern Humans and the Impact of Chronometric Dating*, Princeton: Princeton University Press, pp. 69–83.

Myers, A. A. and Giller, P. S. 1988. *Analytical Biogeography*. London: Chapman and Hall.

Napier, J. R. 1971. *The Roots of Mankind*. London: Allen and Unwin.

Oakley, K. 1959. *Man the Toolmaker*. London: British Museum (Natural History).

Pilbeam, D. 1967. Man's earliest ancestors. *Science Journal*, 3: 47–53.

Pinker, S. 1994. *The Language Instinct*. London: Allen Lane.

Plomin, R., DeFries, J. C. and McClearn, G. E. 1990. *Behavioural Genetics: a Primer*. New York: Freeman.

Popper, K. 1974. *The Philosophy of Karl Popper* (ed. P. A. Schilpp). La Salle, IL: Open Court.

Reader, J. 1988. *Missing Links*. (2nd edition) London: Pelican.

Ridley, M. 1993. *The Red Queen*. New York: Viking.

Rightmire, G. 1990. *The Evolution of Homo erectus*. Cambridge: Cambridge University Press.

Rodman, P. S. 1984. Foraging and social systems of orang utans and chimpanzees. In P. S. Rodman and J. G. H. Cant (eds), *Adaptations for foraging in Non-Human Primates*, New York: Columbia University Press, pp. 134–60.

Rodman, P. S. and McHenry, H. M. 1980. Bioenergetics and origins of bipedalism. *American Journal of Physical Anthropology*, 52: 103–6.

Rodman, P. and Mitani, 1985. Orang utan: sexual dimorphism in a solitary species. In B. B. Smuts, D. L. Cheney, R. M. Seyfarth, R. W. Wrangham and

T. T. Struhsaker (eds), *Primate Societies*, Chicago: University of Chicago Press, pp. 146–54.

Rodspeth, L., Wrangham, R. W., Harrigan, A. M. and Smuts, B. 1991. The human community as a primate society. *Current Anthropology*, 32: 221–54.

Roede, M., Wind, J., Patrick, J. M. and Reynolds, V. (eds) 1991. *The Aquatic Ape: Fact or Fiction.* London: Souvenir Press.

Rudwick, M. J. S. 1971. Uniformity and progression: reflections on the structure of geological theory in the age of Lyell. In D. H. D. Roller (ed.), *Perspectives in the History of Science and Technology*, Norman: Oklahoma University Press, pp. 209–27.

Ruse, M. 1979. *The Darwinian Revolution: Science Red in Tooth and Claw.* Chicago: University of Chicago Press.

Ruse, M. 1986. *Taking Darwin Seriously.* Oxford: Basil Blackwell.

Sarich, V. and Wilson, A. 1967. Immunological timescale for evolution. *Science*, 158: 1200–3.

Schick, K. D. and Toth, N. 1993. *Making Silent Stones Speak.* New York: Simon and Schuster.

Senut, B. and Tardieu, C. 1985. Functional aspects of Plio-Pleistocene hominid limb-bones: implications for taxonomy and phylogeny. In E. Delson (ed.), *Ancestors: The Hard Evidence*, New York: Alan R. Liss, pp. 193–201.

Sibley, C. G. and Alquist, J. E. 1984. The phylogeny of the hominoid primates as indicated by DNA-DNA hybridisation data. *Journal of Molecular Evolution*, 20: 2–15.

Skelton, R. R. and McHenry, H. M. 1992. Evolutionary relationships among early hominids. *Journal of Human Evolution*, 23: 309–49.

Smuts, B. B., Cheney, D. L., Seyfarth, R. M., Wrangham, R. W. and Struhsaker, T. T. (eds) 1985. *Primate Societies.* Chicago: University of Chicago Press.

Spencer, F. (ed.) 1985. *A History of American Physical Anthropology.* New York: Academic Press.

Spencer, F. 1990. *Piltdown, A Scientific Forgery.* London: Natural History Museum.

Spencer, H. 1851. *Essays Scientific, Political and Speculative.* London.

Stanley, S. M. 1978. Chronospecies' longevity, the origin of genera, and the punctuational model of evolution. *Palaeobiology*, 4: 26–40.

Stoneking, M. 1993. DNA and recent human evolution. *Evolutionary Anthropology*, 2(2): 60–73.

Stringer, C. and Gamble, C. 1993. *In Search of the Neanderthals.* London: Thames and Hudson.

Stringer, C. B. and Andrews, P. J. 1988. Genetic and fossil evidence for the origin of modern humans. *Science*, 239: 1263–8.

Swisher, C. C., Curtis, G. H., Jacob, T., Getty, A. G., Suprijo, A. and Widiasmoro, 1994. Age of the earliest known hominids in Java, Indonesia. *Science*, 263: 1118–21.

Symons, D. 1979. *The Evolution of Human Sexuality.* New York: Oxford University Press.

Tattersall, I. 1986. Species recognition in palaeontology. *Journal of Human Evolution,* 15: 165–75.

Templeton, A. R. 1993. The 'Eve' hypothesis: a genetic critique and reanalysis. *American Anthropologist,* 95: 51–72.

Thomson, D'Arcy 1942. *On Growth and Form.* Cambridge: Cambridge University Press.

Tobias, P. V. 1980. A survey and synthesis of the African hominids of the Late Tertiary and early Quaternary periods. In L. K. Konigsson (ed.), *Current Arguments on Early Man,* Oxford: Pergamon, pp. 86–113.

Tobias, P. V. 1991. *Olduvai Gorge, Volume 4: The Skulls, Endocasts and Teeth of Homo Habilis.* Cambridge: Cambridge University Press.

Tooby, J. and Cosmides, L. 1992. Psychological foundations of culture. In J. Barkow, L. Cosmides and J. Tooby (eds), New York: Oxford University Press, pp. 19–136.

Turner, A. and Wood, B. A. 1993. Taxonomic and geographic diversity in robust australopithecines and other Plio-Pleistocene mammals. *Journal of Human Evolution,* 24: 147–68.

Van Daniken, E. 1968. *Chariots of the Gods.* London: Souvenir Press.

von Koeningswald, G. H. R. and Weidenreich, F. 1939. The relationship between *Pithecanthropus* and *Sinanthropus. Nature,* 144: 926–9.

Vrba, E. 1985. Environment and evolution: alternative causes of the temporal distribution of evolutionary events. *South African Journal of Science,* 81: 229–36.

Walker, A. C., Leakey, R. E., Harris, J. M. and Brown, F. H. 1986. 2.5 Myr *Australopithecus boisei* from west of Lake Turkana, Kenya. *Nature,* 322: 517–22.

Walker, A. C. and Leakey, R. E. 1993. *The Nariokotome* Homo erectus *Skeleton.* Cambridge, MA: Harvard University Press.

Wallace, A. R. 1870. *Contributions to the Theory of Natural Selection.* London.

Wallace, A. R. 1889. *Darwinism: An Exposition of the Theory of Natural Selection.* London.

Washburn, S. L. 1983. The evolution of a teacher. *Annual Review of Anthropology,* 12, 1–24.

Weidenreich, F. 1943. The skull of *Sinanthropus pekinensis*: a comparative study on a primitive hominid skull. *Paleontologica sinica,* New Series D.10: 1–291.

Wheeler, P. E. 1985. The evolution of bipedalism and the loss of functional body hair in hominids. *Journal of Human Evolution,* 14: 23–8.

Wheeler, P. E. 1991a. The thermoregulatory advantages of hominid bipedalism in open equatorial environments: the contribution of increased convective heat loss and cutaneous evaporative cooling. *Journal of Human Evolution,* 21: 107–16.

Wheeler, P. E. 1991b. The influence of bipedalism on the energy and water budgets of early hominids. *Journal of Human Evolution,* 21: 117–36.

Wheeler, P. E. and Aiello, L. C. 1995. The expensive tissue hypothesis. *Current Anthropology,* 36: 199–222.

White, T. D., Johanson, D. C. and Kimbel, W. H. (eds), 1983. *Australopithecus africanus:* its phyletic position reconsidered. In R. L. Ciochon and R. S. Corruccini (eds), *New Interpretations of Ape and Human Ancestry,* New York: Plenum, pp. 721–9.

White, T. D., Gen Suwa, Hart, W. K., Walters, R. C., WoldeGabriel. G., De Heinzelin, J., Clark, J. D., Asfaw, B. and Vrba, E. 1993. New discoveries of *Australopithecus* at Maka in Ethiopia. *Nature,* 366: 261–5.

White, T. D., Suwa, G. and Asfaw, B. 1994. *Australopithecus ramidus,* a new species of early hominid from Aramis, Ethiopia. *Nature,* 371: 306–12.

Whiten, A. 1993. *Natural Theories of Mind.* Oxford: Basil Blackwell.

Williams, G. G. 1966. *Adaptation and Natural Selection.* Princeton: Princeton University Press.

Williams, S. A. and Goodman, M. 1989. A statistical test that supports a human/chimpanzee clade based on non-coding DNA sequence data. *Molecular Biology and Evolution,* 6: 325–30.

Wilson, E. O. 1976. *Sociobiology.* Cambridge, MA: Harvard University Press.

WoldeGabriel, G. et al. 1994. Ecological and temporal placement of early Pliocene hominids at Aramis, Ethiopia. *Nature,* 371: 330–3.

Wolpoff, M. H. 1971. Competitive exclusion among lower pleistocene hominids: the single species hypothesis. *Man,* 6: 601–14.

Wood, B. A. 1984. The origins of *Homo erectus. Courier Forschungsinstitut Seckenberg:* 99–111.

Wood, B. A. 1991. *Koobi Fora Research Project, Volume 4: The Hominid Cranial Remains.* Oxford: Clarendon Press.

Wood, B. A. 1994. The oldest hominid yet. *Nature,* 371: 280–1.

Wrangham, R. W. 1980. An ecological model of female-bonded primate groups. *Behaviour,* 75: 262–300.

Wrangham, R. W. 1987. The significance of African apes for reconstructing human evolution. In W. G. Kinzey (ed.), *The Evolution of Human Behavior: Primate Models,* Albany: SUNY Press, pp. 28–47.

Wright, D. H. 1983. Species-energy theory: an extension of species area theory. *Oikos,* 41: 496–506.

Zetterberg, J. P. (ed.) 1983. *Evolution versus Creationism.* Phoenix: Oryx Press.

Index

Acheulian, 75
Activity patterns
  of primates, 144
  and bipedalism, 145–6
Adams, D., 19
Adaptation, 27–8, 121–2, 132–5
Adaptive radiations, 101–3
Africa
  climatic change in, 116
  dispersal from, 120–3
  environments, 139
  as evolutionary centre, 113–14,
    121
  as an evolutionary community,
    113–14
  geology of, 109–13
  and human evolution, 105–31
  and human genetics, 123–31
  and Louis Leakey, 106–7
African Eve, 123–4, 126–30
Aiello, L., 192
Allometry, 160–5
  see also brain size
Anatomy
  comparison of apes and
    humans, 32–9

comparison of archaic and
    modern humans, 124
Andrews, P., 88
Apes
  biogeography of, 107–9
  divergence from monkeys, 182
  evolutionary relationship to
    humans, 62–5
  and humans 32–9
  Machiavellian, 170
  social evolution, 182, 183
Aquatic ape hypothesis, 17, 142
Ardrey, R., 195
Asia
  as centre of human evolution,
    106
  hominids in, 88
  colonization by hominids,
    121–2
Australia, 124
Australopithecus aethiopicus, 84,
  98–100
Australopithecus afarensis, 51, 74,
  80–3, 98–100
Australopithicus africanus, 56, 83,
  98–100

Australopithecus anamensis, 83, 87, 98
Australopithecus boisei, 84, 98–100
Australopithecus crassidens, 84, 98–100
Australopithecus ramidus, 70, 77, 83, 98–100
Australopithecus robustus, 84, 96, 98–100
Awash, 70
Aye-aye, 155

Baldwin, P., 140
Baroba, 14
Beetles, evolutionary abundance of, 150
Behaviour
  evolution of human behaviour, 182–9
  and evolution, 134–5
  see also social behaviour
Benefits
  costs and benefits in evolution, 155, 170–1, 173, 190–1
Biogeography, 62–6, 79, 92, 107–9
Bipedalism, 41, 71, 74, 117–18, 133, 135–8, 143–7, 187–8
Boas, F., 4
Body size, 81
  see also allometry; brain size; hominids
Bovids, 95–6
Brain size, 33, 74–5, 160–5, 170–1, 190
Broom, R., 83
Byrne, R., 167, 169

Cann, R., 128
Catarrhini, 178
Cercopithecoidea, 178
Chimpanzees
  brain size, 162
  close relationship with humans, 62
  foraging behaviour, 139–40
  social behaviour 178–9

Climatic change, 138
Cognition
  evolution of human cognition, 202–5
  evolution and, 158
Cosmides, L., 197
Costs, *see* benefits
Creation myths
  differences from evolution, 16
Cultural relativism, 3–5
Culture, 45–6, 151, 154, 168, 193, 195, 197–200

D'Arcy Thomson, 162
Dart, R., 44, 56, 83
Darwin, C., 16, 29, 34, 52, 196
  and bipedalism, 136
  and chronology, 52–5
  and genetics, 6
  and human uniqueness, 34–8
Darwinian revolution, 1–2, 172
Darwinism, 19–31
  and anthropology, 9
  biological alternatives to, 6–7
  and competition, 9–10
  components of, 24–6
  and creation myths, 14
  and ecology, 9
  and the human biological heritage, 207–12
  and humans, 2, 21–3
  and psychology, 9
  rejection of, 4, 8–9
  resurgance in, 9
  see also evolutionary theory; natural selection
Dawkins, R., 153, 198
Determinism
  biological, 2
  genetic, 2
Dinosaurs, 49, 105
Disraeli, B., 32
Diversity
  in human evolution, 72, 85–7, 92–4
  species diversity, 119–20
Dolphins, 151

Dover, G. 7
Dunbar, R. 144, 145, 148, 170, 198
Durkheim, 3

Ecology
    and abundance, 150
    and brain size, 166–7
    of colonizing species, 122–3
    and evolution, 119
    of males and females, 176,
        180–9
    and social behaviour, 176–82
    and species diversity, 119–20
EEA, see environment of
        evolutionary adaptedness
Eldredge, N., 7
Environment of evolutionary
        adaptedness, 205–7, 207–12
Eugenics, 3
Europe
    colonization by hominids,
        121–2
    hominids in, 88
Evans-Pritchard, E. P., 4
Eve, see African Eve
Evolution
    abundance and rarity in,
        150–5
    and chance, 152
    consequence of natural
        selection, 27
    and creation myths, 14
    criticism of, 1–8, 23–4
    and ideology, 9
    and laws, 118–20, 173
    mutations as a factor, 152
    and neutrality, 66–7
    and progress, 3–4, 94–7
    scale of, 59–60
    and social anthropology, 4
    and social behaviour, 168–70
    and social science, 2–3
    and social theory, 3
    as a survival strategy, 13
Evolutionary heritage for
        humans, 207–11
Evolutionary psychology, 206

Evolutionary rates, 154
Evolutionary theory,
    centre and edge hypothesis,
        123
    changes in, 5
    costs and benefits, 155–60
    historical constraints and,
        115–18
    and human origins, 22– 3 see
        also Darwinism
    testability, 23–4
Exaptation, 47
Exogeny, 122

Fathers, role in human evolution,
        189
Finite social space, see social
        behaviour
Fisher, R. A., 35
Fortes, M., 4

Gellner, E., 197
Genes
    determination and under-
        determination, 197–9
Genitalia
    humans and apes compared,
        41–2
Goodall, J., 39, 46
Gould, S. J., 7, 9, 47, 96

Hair
    loss in humans, 41, 142–3
    loss in mammals, 141–3
Haldane, J. B. S., 150
Hardy, A., 142
Heat stress
    in human evolution, 142
Hinde, R., 169, 200
Hominid evolution, see human
        evolution
Hominids
    as African apes, 107–9
    biogeographic patterns, 107–9
    body weight, 81
    as colonizers, 120–3
    definition of, 79

diversity, 72, 85–7, 92–4
earliest evidence for, 70–1
and foraging behaviour,
    138–41
number of species, 93
phylogeny, 91
social behaviour, 182–9
taxa, 82
Hominoidea, *see* Hominoids
Hominoids, 62
and humans, 78–9
Homo, 99–100
earliest evidence for, 72
Homo erectus, 74, 87–90, 99–100,
    124
Homo ergaster, 87, 99–100
Homo habilis, 84–5, 99–100
Homo rudolfensis, 85, 99–100
Homo sapiens, 88–90, 93, 99–100,
    123–5
Howell, F. C., 93
Hull, D. L. 198
Human evolution
adaptive radiations, 101–3
and brain size, 160–5
chronological scale of, 51
and cognition, 202–5
costs and benefits in, 157–8
and culture, 154
and environmental change, 183
explanations of, 173–4
fossil evidence for, 70–1
geographical basis for, 92
and geological context, 109–13
and kinship, 183–5
and language, 204–5
long and short chronology,
    51–8
modern human origins,
    123–31
molecular evolution, 69
pattern of, 80–92
phylogeny of, 98–101
progress in, 94–7
scale of, 73
significance for modern
    humans, 195–201

and social evolution, 182–9
trends in, 97–8
Humans
and African apes, 62–6
behavioural characteristics,
    75–7
convergence with marine
    mammals, 142
costs and benefits of, 155–9
criteria for, 72–7
definition, 40–6
differences from animals,
    35–8
and natural selection, 16–19
social behaviour, 177
uniqueness among mammals,
    150–5
uniqueness and the fossil
    record, 103–4
Humphrey, N., 167, 169, 190, 202
Hunting, 44–5, 191–2
Huxley, J. 162
Huxley, T. H., 56, 196

Individuals
in evolution, 133–4
Infanticide, 189
Inheritance
mechanisms of , 6
Intelligence, 159, 198
    see also brain size
Invasions, by exogenous species,
    122

Jacob, F., 115
Java, 124
Johanson, D., 71, 80, 81

Keith, A., 53–4, 58
Kenyapithecus, 54
Koobi Fora, 84
Kortlandt, A., 39, 154

Laetoli, 71, 80
Lake Turkana, 70, 84
Lamarckism, 23
Landau, M., 14–15

Language, 45, 151, 158–9, 201–2
Latitude
  and species diversity, 119–20
Leach, E., 3
Leakey, L. S. B., 54, 84, 106
Leakey, M., 71, 84
Leakey, R., 54, 56, 84
Lee, P. C., 169, 176, 177
Lee, R. B., 45
Levi-Straus, C., 3
Lewis, R., 135
Life
  the universe and everything,
    answer to, 19
  see also 42
Life history
  changes in human evolution,
    192–4
Longevity, 193–4
Lothagam, 71
Lovejoy, O., 137, 148
Lyell, C., 52

Male kin-bonding, 178–9
Malinowski, B., 4
Martin, R. D., 192
Maynard Smith, J., 2
Mayr, E., 123
McGrew, W. C., 140
McHenry, H., 81, 140
Medawar, P., 19
Mendel, G., 6
Menopause, 193
Metabolism
  and brain size, 170, 191
Mind, theory of, 203
Mitochondrial DNA, 126–9
  and modern human origins,
    126–30
  see also African Eve
Models
  in human evolution, 46–8
Modern humans origins, 132–1,
  see also Homo sapiens
Molecular drive, 7
Molecular evolution, 65–7
  see also mitochondrial DNA

Monogamy
  and bipedalism, 137
  in social evolution, 177
Morgan, E., 17, 142
Morris, D., 11, 141
Mothers, role in human
    evolution, 189–90
Mousterian, 75
mtDNA, see mitochondrial DNA

Natural selection, 6, 16, 24–6
  and chance, 152–3
  and cultural behaviour, 200
  as a deterministic process,
    153–4
  and explaining human
    evolution, 173, 196
  limits in human evolution,
    35–8
  replicators, interactors and
    vehicles, 198
Nature–nurture debate, 197
Neanderthal, 29, 74, 96, 124
neo-Darwinism, 7
  see also Darwinism
Ngaloba, 124

Oakley, K., 46
Oldowan, 75
Olduvai, 75, 84
Omo Kibbish, 124

Paley, 152
Paranthropines, see robust
    australopithecines
Parental care, 189–90
Phylogeny, 72
  of hominids, 98–101
Piltdown, 56, 106
Pinker, S., 205
Polygamy, 177
Popper, K., 23–4
Primates
  as social specialists, 159–60,
    167, 174–5
Provisioning
  and bipedalism, 137

Punctuated equilibrium, 7

Ramapithecus, 54–5, 68
Rarity
  in evolution, 150–5
Reductionism, 20–1
Religion
  versus science, 1–2
Resources
  and social behaviour, 179
Rift Valley, 70, 80, 110–12
Robust australopithecines, 72, 84,
  92–4
Rodman, P., 140
Rudspeth, L., 177

Science
  and human evolution, 19–20
Senud, B., 81
Sex
  ecological differences between
    males and females, 176
Sexual dimorphism, 42, 193
Sexual selection, 35–8
Social behaviour
  in biology, 169
  definitions, 167–8
  and ecological factors, 176–82
  finite social space model,
    175–6
  among primates, 167, 174–5
  kin-bonded systems, 177
Social Darwinism, 3
Social evolution 169–71, 174–82,
  human social evolution, 182–9
  timetable for, 207–11
Social intelligence, 159–60
Sociality, *see* social behaviour
Species
  in evolution 132–3
  of hominid, 92, 93
  see also hominids
Species diversity, *see* diversity

Spencer, H., 3
Standard Social Science Model,
  197, 206
Stoneking, M., 128
Strategy
  as an evolutionary concept,
    133–5

Tabarin, 71
Tattersall, I., 93
Taxonomy, 60–2
Technology, 43–4, 75, 76
Thermoregulation
  in human evolution, 141–3.
Time
  as an ecological constraint, 144
  and human evolution, 49–51
Tooby, J., 197
Transvaal, 70, 110–12
Tropical rainforests, 119
Tropics
  role in evolution, 119–20

Uniformitarianism, 52

van Daniken, 17
Von Koeningswold, G. H. R., 106
Vrba, E., 47

Walker, A., 84
Wallace, A., 34
Washburn, S., 57
Weber, M., 3
Weidenreich, F., 106
Wheeler, P., 142–3, 146, 192
White, T., 80, 81
Whiten, A., 167, 169
Wilberforce, S., 1
Williams, G. C., 9
Wilson, A., 128
Wolpoff, M., 154
Wood, B., 84–5, 93
Wrangham, R., 176